John R. Eastman

The Progress of Meteoric Astronomy in America

Read before the Philosophical Society of Washington, April 12, 1890

John R. Eastman

The Progress of Meteoric Astronomy in America
Read before the Philosophical Society of Washington, April 12, 1890

ISBN/EAN: 9783337068882

Printed in Europe, USA, Canada, Australia, Japan

Cover: Foto ©berggeist007 / pixelio.de

More available books at **www.hansebooks.com**

PHILOSOPHICAL SOCIETY OF WASHINGTON
BULLETIN VOL. XI, pp. 275-358.

THE PROGRESS

OF

METEORIC ASTRONOMY

IN

AMERICA.

BY

JOHN ROBIE EASTMAN.

READ BEFORE THE PHILOSOPHICAL SOCIETY OF WASHINGTON,
APRIL 12, 1890.

WASHINGTON:
PUBLISHED BY THE SOCIETY
JULY, 1890.

THE PROGRESS OF METEORIC ASTRONOMY IN AMERICA.

BY

John Robie Eastman.

[Read before the Society April 12, 1890.]

TABLE OF CONTENTS.

INTRODUCTION.

The progress of Meteoric Astronomy through its successive stages of development has been so peculiar in America, especially in the United States, that, unlike almost all the other branches of Astronomy and Physics, its advance may be thoroughly discussed with very little reference to the important growth which it has made in Europe.

From the nature of the phenomena it is evident that the apparition and fall of meteors must have compelled the attention of mankind through all ages, but the earliest records are at least obscure. While there may be some claim to authenticity in the early allusions to what was apparently meteoric phenomena, there seem to be no trustworthy observations until about 600 B. C.

From that time the falls of a great number of meteors and meteorites were recorded with more or less accuracy and detail, but no special attention was attracted to the observation and study of such phenomena until the publication of a paper in 1794, by Chladni, on a mass of meteoric iron found in Siberia by Dr. Pallas, a well-known naturalist. About this time several noted meteorites fell in Europe, and in 1802 Edward Howard published in the Philosophical Transactions a paper entitled " Experiments and observations on certain stony substances which at different times are said to have fallen on the earth." This paper contains, probably, the account of the first chemical analysis of a meteorite ever made.

Nearly all the publications referring to meteors, both in Europe and America, up to the year 1833 were confined to vague theories and brief speculations with regard to their origin.

The very important meteoric shower on the morning of April 20, 1803, was the first well-defined phenomenon of that class in this country of which there seems to be any record. There is no evidence that it was well observed except at Portsmouth, N. H., and at Richmond, Va., and no recurrence of this shower of any notable magnitude has since been observed. Graphic accounts of this phenomenon were printed in " The New Hampshire Gazette," of Portsmouth, N. H., May 31, 1803, and in " The Virginia Gazette and General Advertiser," of Richmond, Va., May 23, 1803, but apparently no scientific interest or discussion was developed.

The wonderful display of meteors on the morning of November 14, 1833, which was seen throughout the Atlantic coast of the United States, gave a decided impulse to the study of the subject and suddenly brought the principal American observers into prominence.

The serious study of meteoric phenomena in America may be said to date from this epoch.

The earliest studies immediately developed theories, more or less fantastic, to account for the varied but startling display in the heavens.

The first theories, derived from only a few facts, naturally presented the greatest range of speculation.

As phenomena multiplied, the limits of speculation were

notably contracted, and in 1834 the germ of the true theory of meteoric motion was presented, but not developed.

The most accurate idea of the progress of the science of Meteoric Astronomy can be obtained, without doubt, from an examination of the principal theories.

ABSTRACTS OF THEORIES.

The following abstracts of these theories are presented in chronological order, and in each case the language of the author is employed if practicable.

Probably the first paper printed in this country which advances any theory of the nature or the motion of meteors was written by *Rev. Thomas Clap*, ex-president of Yale College, and was printed in Norwich, Connecticut, in 1781. He concluded that "our observations have heretofore been so imperfect as that we cannot easily determine minute circumstances; but the general theory seems highly probable, if not certain, that these superior meteors are solid bodies, half a mile in diameter, revolving around the earth in long ellipses, their least distance being about twenty or thirty miles; that by their friction upon the atmosphere they make a constant rumbling noise and collect electrical fire, and when they come nearest the earth or a little after, being then overcharged, they make an explosion as loud as a large cannon."

In 1819 *W. G. Reynolds, M. D.*, of Middletown Point, N. J., published a paper[1] advocating the theory that "Meteors proceed from the earth. They arise from certain combinations of its elements with solar heat, and meteoric stones are the necessary results of the decomposition of these combinations."

After the shower of 1833 elaborate accounts of the event were written by several scientific observers, and various conclusions and theories were deduced.

Prof. Edward Hitchcock,[2] of Amherst College, Mass., con-

[1] A. J. S., I₁, 266.   [2] A. J. S., XXV₁, 354.

cluded that "there was a point from which most of the meteors seemed to emanate; that this radiant corresponded to that point in the dome of the heavens to which the magnetic needle would point if left free to move vertically and horizontally, and that meteors are only modifications of the Aurora Borealis."

Prof. D. Olmstead,[1] of Yale College, discussed at length the meteors of November 13, 1833, with the following conclusions:

"1st. The meteors originated beyond the limits of our atmosphere and fell towards the earth, in straight and nearly parallel lines, from a point 2,238 miles above the surface of the earth.

"2d. Their velocity on entering the earth's atmosphere was about four miles per second.

"3d. They consisted of light, transparent, combustible matter, and took fire and were consumed in traversing the atmosphere."

Prof. Olmstead finally concluded that "the meteors of November 13 consisted of portions of the extreme parts of a nebulous body which revolves around the sun in an orbit interior to that of the earth, but little inclined to the plane of the ecliptic, having its aphelion near to the earth's path and having a periodic time of 182 days, nearly."

After discussing the November meteors of 1836, Prof. Olmstead[2] concluded that "the zodiacal light might be the source of those meteors, and therefore was not a portion of the sun's atmosphere, but a nebulous or cometary body revolving around the sun within the earth's orbit nearly in the plane of the solar equator, approaching at times very near to the earth, and having a periodic time of either one year or half a year, nearly."

On the 28th of April, 1840, *Mr. E. C. Herrick*[3] read before the Connecticut Academy of Arts and Sciences a paper on

[1] A. J. S., XXVI₁, 132. [2] A. J. S., XXXI₁, 386.
[3] A. J. S., XL₁, 349.

" The history of star-showers of former times," in which he
presented a brief account of all the records he had been able
to find, together with the following tabular chronological
account of star-showers, where the dates are reduced to Gre-
gorian style:

Chronological List of Star-Showers.

Number.	Date.	Number.	Date.
1	B. C. 1768.	21	A. D. 1060.
2	B. C. 686.	22	" 1090.
3	A. D. 7.	23	" 1094.
4	" 532.	24	" 1095, April 10.
5	" 558.	25	" 1096, April 10 (?).
6	" 585, September 6 (?).	26	" 1106, February 19.
7	" 611.	27	" 1122, April 11.
8	" 744 or 747.	28	" 1199, October (?).
9	" 750.	29	" 1202, October 26.
10	" 764 (?), March.	30	" 1213, August 2.
11	" 765, January 8.	31	" 1366, October 30.
12	" 829.	32	" 1398.
13	" 855, October 21.	33	" 1399, October (?).
14	" 899, November 18.	34	" 1635 or 1636.
15	" 901, November 30.	35	" 1743, October 15.
16	" 902, October 30.	36	" 1799, November 12.
17	" 912 or 913.	37	" 1803, April 20.
18	" 931 or 934, October 19.	38	" 1832, November 13.
19	" 935, October (?).	39	" 1833, November 13.
20	" 1029, July or August.		

Of the theory of meteors, Mr. Herrick wrote:

"The most probable hypothesis is that there are revolving around the sun millions of small planetary and nebulous bodies of various magnitudes and densities, and that when any of these dart through our atmosphere they become ignited and are seen as shooting-stars."

In discussing a paper on meteors by Prof. Erman (Schumacher's Ast. Nach. No. 385) *Prof. Benjamin Peirce*,[1] after pointing out an error in Erman's work, concludes in these words: "The plane of the meteors cannot differ much from that of the ecliptic, and their relative velocity cannot exceed one-third of the earth's velocity. A ring so nearly in the plane of the earth's orbit must be subject to great perturbations; and, if there is one, I think that no observations which we can make will enable us to calculate its motions with any degree of accuracy."

On January 15, 1841, *Prof. S. C. Walker*[2] read a paper before the American Philosophical Society, " On the periodical meteors of August and November," in which the following points were discussed :

The relative velocities of meteors ;

The relative directions of meteors in space ;

The periodical or anniversary display of meteors ;

The respective plausibilities of the hypotheses of a single cluster with a half-yearly or yearly period, or that of a continuous ring for the periodical meteors of August and November ;

The theories of aerolites and shooting-stars ;

The variation of the relative velocity and of the convergent point ;

And principally the investigation of formulæ for computing the elliptic elements of the orbit of a meteor from its observed relative velocity and direction.

In 1844 an " Essay on Solid Meteors and Meteoric Stones "

---

[1] Trans. Am. Phil. Soc., VIII$_2$, 83.

[2] Trans. Am. Phil. Soc., VIII$_2$, 87.

was published by *Prof. Peter A. Browne*, of La Fayette College.

The author devoted the first and larger portion of his paper to proving the solidity of meteors.

The latter portion of his essay was confined to the examination and rejection of all the theories previously advanced, which, briefly stated, were :

1st. Dr. Halley's theory that meteors were nothing but a stratum of inflammable vapor, gradually raised from the earth and accumulated in an elevated region, which suddenly took fire at one end and the progress of the flame along the stratum produced the apparent motion of the meteor.

2d. The theory in Luke Howard's Meteorology that hydrogen gas dissolves various bodies, even iron, and that is evolved, mixed with carbon in the gaseous state, from the earth in large quantities, is collected in vast fields in the air, is fired by electric explosions, and, the gasses burning out, they let fall the earthy and metallic contents precipitated and agglutinated as we find them in aerolites.

3d. Prof. Soldani's theory that meteoric stones are generated in the air by a combination of mineral substances which had risen as exhalations from the earth.

4th. Dr. Reynolds' theory, previously given in this paper.

5th. Dr. Blagden's theory that meteors are electrical phenomena.

6th. The theory of Patrick Murray that meteors originate in the local atmosphere of the earth, and their explosions are due to electrical action.

7th. The theories of Brewster and La Grange that meteors are bodies thrown off from the earth by volcanic action.

8th. The theories of Hutton and La Place that meteors are thrown from volcanoes in the moon.

9th. Newton's theory that they proceed from the tail of a comet.

10th. They are terrestrial comets—a theory maintained by Professors Clap and Day and by Carvallo.

11th. The theory that they were solids that have been floating in space from the beginning; advocated by Chladni, Franklin, and Rittenhouse.

12th. The theory of Olbers that they are fragments of an exploded planet.

13th. The theory of Quetelet that they belong to a zone through which the earth passes annually.

14th. The theory of Boubée that they are fragments of an exploded comet.

These theories are all rejected as disproved or absurd; but the author advances no theory as a substitute. He announces, however, that, to his mind, "the most probable supposition yet made is that the solid meteors may possibly emanate from the sun," though no serious attempt is made to prove the proposition.

In a paper read by *Prof. J. Lawrence Smith* before the American Association for the Advancement of Science, in April, 1854, the author[1] advocated the theory of the lunar origin of meteors, which he stated as follows: "The moon is the only large body in space, of which we have any knowledge, possessing the requisite conditions demanded by the physical and chemical properties of meteorites; and they have been thrown off from that body by volcanic action (doubtless long since extinct), and, encountering no gaseous medium of resistance, reached such a distance as that the moon exercised no longer a preponderating attraction, the detached fragment possessing an orbital motion and an orbital velocity which it had in common with all parts of the moon, but now more or less modified by the projectile force and new condition of attraction in which it was placed in reference to the earth, acquired an independent orbit more or less elliptical. This orbit, necessarily subject to great disturbing influences, may sooner or later cross our atmosphere and be intercepted by the body of the globe."

In 1859[2] *Dr. B. A. Gould* read a paper before the Ameri-

---

[1] A. J. S., XIX₂, 343.   [2] Proc. A. A. A. S. 1859, 181.

35—Bull. Phil. Soc. Wash., Vol. II.

can Association for the Advancement of Science to disprove
the theory that meteors had their origin in lunar valcanoes.

Assuming that a lunar volcano may eject masses of matter
with the requisite velocity to pass beyond the region where
the lunar gravitation predominates over the terrestrial and
the masses become obedient to the earth's attraction, Dr.
Gould examines in detail the consequences to which the
theory of the lunar origin of meteorites would necessarily
lead, and presents his conclusions as follows:

"From the foregoing considerations we are warranted in
assuming that for every body expelled from lunar volcanoes
with a force adapted for throwing to the earth an acrolite of
average dimensions there are, in probability, at the very
least one hundred and eighty bodies expelled with forces
not thus adapted; that for every mass ejected with the average force of a lunar volcano and striking the earth, there
are at least one hundred and eighty masses of inadequate
dimensions ejected; and that for any given combination of
volcanic force and projected mass, the region of the lunar
surface, whence the mass may reach the earth, is exceeded
in extent by the tract of the lunar surface whence this would
be impossible, in the ratio of fifty to one. Combining these
several individual probabilities, it will readily be perceived
that not more than

$$\tfrac{1}{50} \times \tfrac{1}{180} \times \tfrac{1}{180} = \tfrac{1}{1620000}$$

of all the ejected lava masses, or about 3 in 5,000,000 of each
possible size, would probably ever reach the earth as aerolites; nor is this an unsafe estimate. It is a very guarded
one, and the fraction $\tfrac{1}{2000000}$ would be more likely to be
correct. Now, can we regard it as probable that the moon
has parted with so large an amount of matter as nearly, if
not quite, two million times the combined mass of all the
aerolites which have fallen to the earth? I think not.
Even of those known to have fallen there are more than five
hundred of various weights, one of them having a mass of
thirty thousand pounds. The tokens of such a mass of
gravitative matter as this would imply could not fail to be

legibly inscribed in the unerring and enduring records of our system. They would preclude the accordance which is found to exist between the present lunar theory and the ancient observations. They would be found to be inconsistent with the known values of precession and nutation. They might, indeed, almost be said to be incompatible with the present mass of the moon."

From the observations of the meteor of November 15, 1859, *Prof. H. A. Newton*[1], of Yale College, concluded that it must have moved in a hyperbolic orbit, and that we have, therefore, two sources of meteors—the solar system and stellar space.

With regard to the periodic meteors of August, *Mr. A. C. Twining*[2], of New Haven, concluded that "the radiant is probably capable of a far more exact determination than is ordinarily supposed or than could have been anticipated, and it is apparently subject to a motion of several degrees from day to day, and a motion which exhibits some remarkable points of agreement in the comparison of one year's positions with those of other years."

From a discussion of the peculiar characteristics of the August meteors *Prof. H. A. Newton*[3] came to the following conclusions:

1st. The individual meteors are cosmical bodies.

2d. They are permanent members of the solar system, revolving about the sun in elliptic orbits.

3d. The direction and velocity of the relative motion, and therefore of the absolute motion of the individual bodies, are nearly the same.

4th. The whole group forms what may be considered a ring or disk around the sun.

5th. The periodic time is two hundred and eighty-one days.

---

[1] A. J. S., XXX₂, 186.   [2] A. J. S., XXXII₂, 444.
[3] A. J. S., XXXII₂, 448.

In the same paper Professor Newton estimates the whole number of meteors in the August ring as 300,000,000,000,000.

In March, 1862, *A. C. Twining*[1], published a paper entitled "Investigations respecting the phenomena of meteoric rings as affected by the earth," and arrived at the following conclusions: "The position of the node of the ring cannot be shifted by the earth's action more than a degree or two in half a million of years; there is an appreciable change of radiant positions, relative to locality on the earth's surface and to the hour of the day, whose maximum is about $3\frac{1}{4}°$ between the extremes and to which the extremes approach; the terrestrial disturbance is sufficient to affect the perihelion distance of the meteors by many millions of miles and to expand the ring to a corresponding breadth at the ascending node; also to collect together in orbits, of similar elements, those meteors which are similarly affected in respect of radiant positions; and terrestrial disturbances do not appear sufficient to draw off meteors into permanently erratic orbits; so that, unless in exceptional instances, meteors are not lost to the ring other than those which the atmosphere absorbs or arrests. If meteors are partially arrested without being dissipated in an excessively tenuous upper medium it may be possible that the ordinary and unconformable meteors are such as have missed a return to the ring under the effect of atmospheric retardation."

Mr. Twining appends the suggestion that, "perhaps comets whose vastly extended atmospheres or heads around the nucleus, although greatly attenuated are perhaps competent to arrest meteors completely, may be found in rare instances to have been disturbed by *impact* with a meteoric ring whose mere attractive influence it would not be possible to detect."

In 1863, *Mr. B. V. Marsh*,[2] of Philadelphia, published a paper on "The luminosity of meteors as affected by latent heat," in which he arrived at the following results: "The upper regions of the atmosphere, even to its utmost limit,

---

[1] A. J. S., XXXIII., 244.   [2] A. J. S., XXXVI., 92.

are grand reservoirs of latent heat most admirably adapted to the protection of the earth from collision with bodies approaching it with planetary velocity from without. The intruder is instantly surrounded with a fiery envelope heated to the greatest conceivable intensity; its surface is burned off or dissipated|into vapor; the sudden expansion of the stratum immediately beneath the burning surface tears the body into fragments, each of which, retaining its planetary velocity, is instantly surrounded by a similar envelope, which produces like effects, and so on until, in most cases, the whole is burned up or vaporized." A second paper on the same subject, and of similar import, was published by Mr. Marsh in the Proceedings of the American Philosophical Society, vol. XIV, 114.

From an examination of the list of November meteor showers from A. D. 902 to A. D. 1833 *Prof. H. A. Newton*[1] concluded that "the star-shower has a motion along the sidereal year of one day in seventy years, and also that the shower has a period of about a third of a century. This precession seems to imply that the orbit of the body furnishing these meteors has only a small inclination to the ecliptic, and that the motion is retrograde. The small distance of the radiant from the point to which the earth is moving, viz., 7°, confirms this conclusion."

In an article on the peculiarities of the November meteors, *Prof. H. A. Newton*[2] arrived at the following conclusions "The length of the annual period as determined from the showers in A. D. 902 and 1833, reckoning 233 leap years, 19 odd days, and adding six hours for difference of longitude, is $365 + \left(\frac{233 + 19.25}{981}\right)$, or 365.271 days. The length of the cycle is 33.25 years.

"The length of the part of a cycle during which showers may be expected may be five or six years or, for extraordinary displays, at least 2.25 years. The supposition of a ring of uniform density throughout its circuit seems improbable.

[1] A. J. S., XXXVI₂, 300. [2] A. J. S., XXXVIII₂, 53.

"The elements of the mean of the orbits of the different groups composing the partial ring are:

$$\text{Semi-major axis} = 0.98049,$$
$$\text{Inclination} = 17°,$$

and the ring is nearly circular.

"The velocity with which these bodies enter the earth's atmosphere is about 20.17 English miles per second."

The most elaborate American paper on meteors up to the date of its publication was prepared in 1865 by *Prof. H. A. Newton*,[1] who discussed the subject under the following divisions, the conclusions being briefly stated in each case:

"1st. The average altitude of the middle points of the luminous portions of the meteor paths is found to be 59.4 English miles.

"2d. The relative frequency of meteors when the heavens were divided into eight equal parts was about equal in all—perhaps a slight preponderance in the southeast—and the relative frequency in different parts of the visible heavens may be considered a function of the zenith distance only.

"3d. Not quite *one* in fifty of all the meteors seen at any one place should have the middle points of their apparent paths within 10° of the zenith.

"4th. The number of visible meteors that come into the atmosphere every day would be 10,460 times the number visible at one station; or the average number that traverse the atmosphere daily, that are large enough to be seen with the naked eye, if the sun, moon, and clouds would permit, would be 30 × 24 × 10,460 = 7,531,000.

"5th. The number of meteoroids in the space which the earth traverses is discussed at length and formulæ are derived for computing the whole number when the average number is known for a given unit of time.

"6th. The average length of the apparent paths derived from 213 European and 803 American observations is 12°.6.

"7th. Adopting the theory that for every meteor visible to

---

[1] Mem. Nat. Acad. Sciences I, 291.

the naked eye there are 52.7 that are visible through a comet-seeker, the whole number of meteoroids coming daily into the air is 400,000,000.

"8th. The mean distance of the meteors from the observer is less than 144 miles.

"9th. The mean foreshortening of the meteor paths by perspective is from 16°.0 to 12°.6.

"10th. The average length of the visible part of meteor paths is between 24 and 40 or 21 and 34 miles; probably nearer 21 and 34 miles.

"11th. The mean duration of flight is not much, if any, greater than half a second of time."

Prof. H. A. Newton[1] discussed the observations of the altitudes of *seventy-eight* meteors observed on November 13–14, 1863, at Washington, Haverford College, Germantown, Philadelphia, and other points, giving diagrams exhibiting the altitudes of these meteors, and also of *thirty-nine* meteors observed in August, 1863, with the following results:

	November meteors.	August meteors.
Mean altitude at appearance	96.2 miles.	69.9 miles.
Mean altitude at disappearance	60.8 "	56.0 "
Mean altitude of middle point of path	78.5 "	62.9 "

In a paper on "The Theory of Meteors" *Prof. Daniel Kirkwood*[2] arrived at the following conclusions:

"The zodiacal light is probably a dense meteoric ring, or rather, perhaps, a number of rings.

"Variable and temporary stars are caused by the interposition of meteoric rings.

"Mercury's mean motion is probably diminished by the action of meteoric matter.

"The transit of a meteoric stream or cloud affords the most probable explanation of the phenomenon known as 'dark days.'

---

[1] A. J. S., XL., 250.   [2] Proc. A. A. A. S. 1866, 8.

" It seems probable that a ring of meteor asteroids exists within the orbit of Titan, Saturn's largest satellite, and causes the annual motion of the apsides of Titan, found by Bessel to be 30' 28".

"Saturn's rings are probably composed of an indefinite number of extremely minute asteroids or meteorites.

" The gaps in the distribution of the mean distances of the asteroids between Mars and Jupiter are analogous to the gaps in Saturn's rings."

In May, 1867, *Prof. Daniel Kirkwood* published a book [1] under the title "Meteoric Astronomy," designed by the author to present in a popular form the principal results of observation and study in that branch of Astronomy. It was devoted chiefly to the collection of some of the principal theories and the more important observations, and to presenting them in a brief but popular form without attempting to set forth any new theory.

A paper by *Prof. H. A. Newton* [2] in 1867, " On certain recent foreign contributions to Astro-meteorology," was devoted to the discussion of a table comparing the epochs and positions of radiant points of shooting-stars concluded independently by R. P. Greg and Dr. E. Heis; the influence of the August and November meteors on the temperature of the atmosphere; the paths and probable origin of the shooting-stars, by Schiaparelli, and the age of the November group of shooting-stars.

From the data obtained from the observations of the November meteors in 1867 *Prof. H. A. Newton* [3] discussed the geographical limits of the shower; the personal equation of observers; the form of the curve of intensity; the breadth of the radiant in latitude; the length of the radiant in longitude, and the distribution in longitude of the perihelia of the orbits of the meteors.

---

[1] J. B. Lippincott & Co., Phila., 1867, 129 pp.
[2] A. J. S., XLIII₂, 285.
[3] A. J. S., XLV₂, 89.

From data obtained from the observations of the November meteors of 1867 at the U. S. Naval Observatory and at Richmond, Va., *Prof. S. Newcomb*[1], U. S. N., computed the altitude of *nine* meteors, finding the mean altitude at apparition to be 102 miles, and at disappearance 47 miles. From data obtained from observations on the same occasion, *Prof. W. Harkness*[2], U. S. N., discussed a method of determining the mass of such meteors as are consumed before reaching the earth.

Assuming that the light produced is always proportional to the amount of material consumed, he arrived at the conclusion that "the mass of the ordinary shooting-stars does not differ greatly from one grain."

In 1869 a paper by *Prof. Daniel Kirkwood*[3] on "Comets and Meteors" was devoted to exhibiting the probable coincidences between the orbits of comets and periodical meteors.

In 1871 *Mr. Jacob Ennis*[4] published a paper entitled "The meteors and their long-enduring trails."

The scope and method of this paper are briefly sketched by the author, and are best presented in his own words, as follows:

"Firstly, I will bring forward many facts to prove that some meteors undergo a process of burning or oxidation while passing through the air, and that the trails are the smoke and ashes of such burning.

"Secondly, I will give facts and reasoning which show that some meteors are composed of various simple chemical elements unoxidized, and which are therefore capable of burning in the air.

"Thirdly, I will show the order and process of creation by which such meteors were originally formed and left in an unoxidized condition."

These points are discussed at length, and numerous theories and observations are cited as proof.

[1] A. J. S., XLV₂, 233.   [2] A. J. S., XLV₂, 237.

[3] Proc. Am. Phil. Soc., XI, 215.   [4] Proc. A. A. A. S. 1871, 122.

In a paper on the "Influence of meteoric showers on auroras" *Prof. Pliny E. Chase*[1] concludes that "there seems therefore good reason to look for an increase of auroral displays soon after every meteoric shower."

In discussing the meteors of November 27, 1872, *Prof. H. A. Newton*[2] remarked, "With Professor Weis and others, I am inclined to consider them all to have been once connected with periodic comets. The scattering took place apparently at or near the perihelion."

In 1872 *Prof. J. W. Mallet*, of the University of Virginia, read a paper[3] on "The occluded gases of meteorites," and another paper[4] by this author on the same subject appeared in 1875.

In 1875 *Prof. A. W. Wright*, of Yale College, published an account[5] of some very carefully conducted experiments made to determine the character and quantity of the occluded gases of meteorites.

From these experiments he derived results differing materially from those obtained by other investigators.

This paper was followed by three others[6] during 1875 and 1876, in which *Professor Wright* reached the conclusion that the spectra of gases from meteorites were identical with the spectra of comets.

In a lecture[7] at the Sheffield Scientific School of Yale College, on "The relation of Meteorites and Comets," *Prof. H. A. Newton*, exhibiting a fragment from the meteoric stone which fell in Iowa February 12, 1875. very clearly presented his theory of the connection of these bodies.

The principal points in the theory, together with some of the arguments, may be briefly stated as follows:

Between the largest meteorite known and the faintest shooting-star that can be seen on a clear night with a telescope

[1] Proc. Am. Phil. Soc. XII, 401.   [2] A. J. S., V₂, 62.
[3] Proc. Royal Society, XX, 365.   [4] A. J. S., X₂, 206.
[5] A. J. S., IX₃, 294.   [6] A. J. S., X₃, 44; XI₃, 253; XII₃, 165.
[7] Nature, Vol. XIX, 315, 340.

there is no essential difference as to astronomical character. In all their characteristic phenomena there is a regular gradation of meteors from one end of the line to the other. They differ in bigness, but in their astronomical relations we cannot divide them into groups. They are all similar members of the solar system. In proof of these statements we cite some of the points in which the large and small meteors are alike and unlike:

1st. They are all solid bodies. It is doubtful whether a small gaseous mass could exist permanently as a separate body in the solar system. A liquid would probably freeze and become solid. In any case, neither a gas nor a liquid could for an instant sustain the resisting pressure which a meteor is subjected to in the air, much less could it travel against it with the velocity observed in ordinary meteor flights. In short, every shooting-star must be a solid body.

2d. The large meteors and the small ones are seen at about the same height from the earth's surface. The air is a shield to protect the earth from an otherwise intolerable bombarding by these meteors. Some of the larger masses penetrate this shield, or, at least, are not melted before their final explosion, when the fragments, their velocity all gone, fall quietly to the ground. The small ones burn up altogether or are scattered into dust.

3d. The velocities of the large and the small meteors agree, and, though they are never measured directly very exactly, we are sure that in general they are more than two and less than forty miles per second. Velocities of from ten to forty miles per second imply that these masses are bodies that move about the sun as a center or else move through space. These velocities, as well as other facts, are utterly inconsistent with a permanent motion of such bodies about the earth or with a terrestrial or a lunar origin.

4th. The motions of the large and small meteors as they cross the sky have no special relations to the ecliptic. If either kind had special relations to the planets, in their origin or in their motions, we should have reason to expect them,

if not always, at least in general, to move across the sky
away from the ecliptic. The fact is otherwise. Both large
and small meteors are seen moving *towards* the ecliptic as
often as *from* it. Neither class seem, therefore, to have any
relation to the planets.

Again, in general character the two classes are alike.
They have like varieties of color; they have similar luminous
trains behind them. In short, we cannot draw any line divid-
ing the stone or iron producing meteor from. the shooting-
star, at least in their astronomical relations. They are all
astronomically alike. They differ in size; but that has nothing to do with their motion about the sun or in space.

The general connection between comets and meteors may
be exhibited in the peculiar relations existing between the
meteors of November 13–14 and their accompanying comet.
The orbit of these meteors is one that is described in 33.25
years. The meteors go out a little further than the planet
Uranus, or about twenty times as far as the earth is from
the sun. While they all describe nearly the same orbit they
are not collected in one compact group. On the contrary,
they take four or five years to pass a given place in the orbit,
and are to be thought of as a train several hundred millions
of miles long but only a few thousands of miles in thickness.

Along with this train of meteors travels a comet. It passed
the place where we meet the meteor stream nearly a year
before the great shower of 1866 and two or three years before
the quite considerable displays of 1867 and 1868. How
came it that this comet and the meteors travel the same
road? The plane of the comet's orbit might have cut the
earth's orbit to correspond with any other day of the year
than November 15; or, cutting it at this place, the comet
might have gone nearer to the sun or farther away; or,
satisfying these two conditions, it might have made any
angle from 0° to 180° instead of 167°; or, satisfying all
these, it might have had any other periodic time than 33.25
years; even then it might have gone off in any other direction of the plane than that in which the meteoroids were

traveling. All these things did not happen by chance; there is something common.

The comet alluded to is not the only one that has an orbit common with meteors, though it is the only case in which the orbit of the meteors is *completely* known, aside from our knowledge of that of the comet. Every August, about the tenth day, we have an unusual number of meteors—a star-sprinkle as it has been called. A comet whose period is about 125 years moves in the plane and probably in a like orbit with these meteors. Near the first of December there have been several star-showers, notably one in 1872, and these meteors are traveling nearly in the orbit of Biela's comet. In April, too, some showers have occurred which are thought to have had something to do with a known comet. Thus much as to the meteors of the star-showers. The sporadic meteors are with good reason presumed to be (and observed facts *prove* some of them to be) the outliers of a large number of meteor streams.

Considering again the November meteor stream and its comet we find that the several bodies move along a common path not at all by reason of a present physical connection. They are too far apart—in general, a thousand times too far apart—to act on each other so much that we may measure the effect. Their connection has been in the past. They must have had some common history. Looking now at the comets, we see that they have been apparently growing smaller at successive returns. Halley's comet was much brighter in its earlier than in its later approaches to the sun. Biela's comet has divided into two or more principal parts, and seems to have entirely gone to pieces. Several comets have had double or multiple nuclei. In the year 1366, in the week after the star-shower, a comet crossed the sky exactly in the track of the meteors. A second comet followed in the same path a week after. Both belonged, no doubt, to the November stream, and one of them may perhaps have been the comet of 1866.

The November meteor stream is a long, thin one. We

have crossed the stream at many places along a length of a thousand millions of miles, sometimes in advance of and sometimes behind the comet, and all along this length have been found fragments—sometimes few, sometimes many. This form of the stream suggests continuous action producing it. A brief, violent action *might* have given this form, but a slowly acting cause seems more natural.

Again, in the history of Biela's comet we have distinct evidence of continued action. The comet divided into two parts not long before 1845, and yet in 1798 fragments of it were met with so far from the comet that they must have left the comet long before, probably many centuries ago.

"Thus we are led to say, *first*, that the periodic meteors of November, of August, of April, &c., are caused by solid fragments of certain known or unknown comets coming into our air; *secondly*, that the sporadic meteors, such as we can see any clear night, are the like fragments of other comets; *thirdly*, that the large fire-balls are only larger fragments of the same kind; and, *finally*, that a portion broken off from one of those large fragments in coming through the air must once have been *a part of a comet*."

"How came the comet to break up? Perhaps the prior question would be, How came the comet together? In its history there is much that cannot yet be explained, much about which we can only speculate."

"Thus, how came this meteoric stone to have its curious interior structure? As a mineral it resembles more the deepest fire-rocks than it does the outer crust of the earth. It seems to have been formed in some large mass, possibly in one larger than any of our existing comets. Some facts show that the comets have almost surely come to us from the stellar spaces. Out somewhere in the cold of space a condensing mass furnished heat for the making of this stone. The surrounding atmosphere was unlike ours, since some of these minerals could hardly have been made in the presence of the oxygen of our air. Either in cooling or by some catastrophe the rocky mass may have been broken to pieces, so as to enter the solar system having little or no cohesion, like a mass of pebbles; or it may have come, and probably did come, a single solid stone.

In either case, as it got near to the sun new and strong forces acted on it.

The same heat and repulsion that develops and drives off from a comet in one direction a tail, sometimes a hundred millions of miles long, may have cracked off and scattered in another direction solid fragments. One of these contained in it this stone, and it wandered in its own orbit about the sun, itself an infinitesimal comet, how many thousands of millions of years we know not, until three years ago it came crashing through the air to the earth in Iowa."

More than ordinary space has been given to the citations from the various statements and arguments and to the concluding speculation of Professor Newton's paper because, better than any preceding American discussion, it presents the status of the modern theories of meteors and comets which are now generally accepted by the scientific world.

The latest formal discussion of this subject was presented by *Professor Newton*[1] in his presidential address before the American Association for the Advancement of Science, at Buffalo, in 1886. This address was devoted wholly to the consideration of the various theories in regard to the motions, character, and origin of meteorites, meteors, and shooting-stars.

The discussion in this address follows the same general lines as in the lecture just cited, while the various arguments are presented with far greater elaboration. No new hypotheses or theories are offered; but the key-note of the address, given in the author's own words, is that "science may be advanced by rejecting bad hypotheses as well as by forming good ones."

---

[1] Proc. Am. Ass. Ad. Science 1886, 1.

EXAMINATION OF THEORIES.

The abstracts and excerpts just presented are, from the limitations of the method employed, frequently very brief, sometimes disconnected, and generally separated from the various discussions which led to the results cited.

Although they present in themselves insufficient data for an accurate study or a rigorous discussion of the subject, they are quite sufficient to illustrate the evolution of the modern theories as they have been successively developed from the superstitions and the dogmatic assumptions of the last century.

This development is a fair illustration of the growth of most of the sciences, and the sometimes absurd and baseless theories, some of which have been cited, are the usual evidences of an anxious, persistent searching after the truth which is satisfied only by success.

While the modern theories have been slowly evolved from a multitude of observations and discussions, expanding here and there along the lines of least difficulty, it is not improbable that frequently there has been a lack of the nicest discrimination as to what were real and well-established facts.

Keeping in view the precept that no sound theory can be based on doubtful data, it is proposed to examine briefly the accumulated mass of so-called knowledge of Meteors and Comets, with a view to ascertaining what we actually *know* about these bodies; what we *infer, assume,* and *assert,* and to some extent, perhaps, what we *do not know* about them.

METEORS.

Those bodies which are usually designated as *meteors, meteorites,* and *shooting-stars* are known, to some extent, by every intelligent person. The first name is usually applied to those sporadic bodies which one can see occasionally on any clear night; the second term is applied to iron or stony masses that sometimes fall to the earth, while the last term is used to designate those bodies which appear in such periodic showers as those of November 13-14, August 6-10, etc., but which, like the first named, are, almost without exception, entirely consumed before they reach the earth.

These bodies have received, at various times, a great variety of names, such as "Fiery Tears of St. Lawrence," "Fire-balls," "Bolides," "Aerolites,' "Meteoroids," etc., most of which have been coined to suit the fancy or ambition of some aspiring author. The only definite knowledge we have of this class of bodies before they reach the surface of the earth is obtained with the spectroscope, and the results from observations with that instrument indicate that all these bodies are similar in composition, and their spectra are the same as that obtained from those masses that have reached the surface of the earth before destruction.

There appears to be, therefore, no reason for using but two names—the one, *meteor*, for those bodies that are consumed before they reach the earth; and the other, *meteorite*, for the solid iron or stony substances that succeed in storming our atmospheric barriers, reaching the surface of the earth intact and bringing our only material messages from the depths beyond.

Sporadic meteors as well as meteorites move apparently in all directions. Meteors that appear in showers seem to emanate from pretty well defined points in the heavens, each separate shower having its own radiant, and in most cases the bodies are not condensed in a single compact mass, but are scattered along the orbit in which they move.

This orbit has been determined for several of the showers with considerable accuracy.

From the testimony of the meteors themselves nothing is known of their origin. The theories of a terrestrial or a lunar volcanic origin are easily shown to be absurd, while the so-called theories that place their origin in other portions of the solar system are mere idle speculations.

COMETS.

The whole number of comets, real and suspected, from about 1770 B. C. to the end of 1889 A. D., the elements of whose orbits have *not* been computed, is 472. From 370 B. C. to the end of 1889 A. D. the number of comets the elements of whose orbits have been computed is 309. Of these, 18 are known to have *elliptic* orbits. In the case of 52, the computed elliptic orbits have not been verified by observation.

The computations show that 231 have *parabolic* orbits, and indicate that 7 have *hyperbolic* orbits.

Thus it appears that not more than seven per cent. of the comets whose orbits have been discussed are known to have *elliptic* orbits, while it is almost certain that seventy-five per cent. have *parabolic* orbits. Of course, the periodic comets, whatever their origin, belong now to the solar system.

As it is highly improbable that there are two or more kinds of comets of intrinsically diverse character and of different origin, it follows that all the comets had their genesis beyond the limits of the solar system, and that the few periodic comets are the exception to the general law, and at best are only adopted members of the solar family.

There are only two sources of actual knowledge of the physical constitution of comets:

One is from the use of the spectroscope; the other is the behavior of the light from a star when seen through various portions, but especially the nucleus of a comet.

As is well known, observations with the spectroscope are not always easily interpreted, but in this case the difficulty is not so great as at first it seems to be.

It is a general law that where there is a continuous spectrum containing all the primary colors without gaps, the light is derived from an incandescent solid or liquid body. A discontinuous spectrum containing bands or bright lines indicates that the light comes from luminous gases or vapors.

To these general rules there are some important exceptions or modifications.

If the temperature of certain vapors or gases be raised to a high degree the number and the appearance of the colored band or of the bright lines change rapidly, though not uniformly, and some investigators have asserted that if the temperature be raised to something over 4,500° Fah. the spectrum will become practically continuous. Similar changes in the phenomena are observed if a gas, like hydrogen, is rendered luminous by the electric spark and then subjected to varying pressures. With a pressure amounting to one-twentieth of an inch of mercury the spectrum is discontinuous and consists of several groups of bright lines in the green. As the pressure is gradually increased there appears a temporary spectrum of bands, then a spectrum of three lines, afterwards a more permanent and complete spectrum of bands, and finally, under a pressure of 52 inches of mercury, a complete and pure continuous spectrum.

The spectra of comets, which have been obtained by careful and experienced observers, present a large number of variations and combinations, ranging from one or more faint bands with indistinct or fluted borders against a colorless background to a faint continuous spectrum with bands or lines of a greater or less degree of brightness and definition.

The most obvious interpretation of the spectroscopic observations of comets is that the bands and lines are the true spectra of a gaseous body, varying through a wide range under the effect of changing pressure and temperature, superimposed upon the faint continuous spectrum derived from the sunlight reflected from the nucleus or other parts of the comet.

Such is the information derived from the spectroscope.

In the vast number of observations of comets, made for the determination of their positions or their physical peculiarities, it has sometimes been noted that the comet passed between the observer and a star without diminishing the apparent brightness of the star or changing its position.

While observing Comet I, 1866, in January, 1866, I saw on one occasion the nucleus of the comet pass directly over a star of the 9.2 magnitude with no more effect on the brightness of the star than would be produced by the close proximity of any object as bright as the comet's nucleus. Similar accounts have been given by other observers, and the phenomenon is too well attested to admit of a reasonable doubt.

The light from a star could not pass unobstructed through a solid body or a dense aggregation of solid bodies; and, considering this phenomenon alone or in connection with the appearance of the nucleus as it approaches and recedes from perihelion, it appears that we are driven to the conclusion that the nucleus of a comet is composed principally, if not entirely, of gaseous matter, which varies in form and in density from the effect of the sun's attraction and repulsion.

COMETS AND METEORS.

The elements of the orbits of *four* meteor streams have been determined with considerable accuracy. These are the streams that produce the showers of November 13–14, November 27, April 20, and August 10.

It has also been found that the orbit of the meteor stream of November 13–14 coincides very closely with the orbit of Comet I, 1866. The orbit of the November 27 stream corresponds to that of Biela's comet, the orbit of the April stream to that of comet I, 1861, and the orbit of the August stream is nearly identical with that of Comet III, 1862.

The identify of these orbits is quite as good as could be expected from the uncertain character of the observations on which the adopted positions of the meteor streams depend.

On these coincidences in the orbits of meteor streams and of certain comets depends principally the modern theory of comets and meteors, which, briefly stated, is as follows:

Sporadic meteors, individual members of meteoric showers,

and meteorites differ in magnitude and appear under widely varying conditions, but from an astronomical standpoint they are all alike.

They are all solid bodies and are fragments of comets.

Assuming that this theory is true, we shall find that some of the inferences drawn from it are of great importance in their bearing on cosmical physics.

1st. As the meteoric masses, both great and small, are derived from comets, they must have originated beyond the limits of the solar system, and they furnish evidence of the existence in space of exactly such minerals, though in different combinations, as are found in the earth's crust.

2d. They arise from the disintegration of comets, which for centuries have furnished the material for the enormous areas of bodies forming the various meteor streams that trail along the orbits of these masses for immense distances.

3d. The meteors forming the shower of November 13–14 have been observed for more than 900 years, and yet the comet whose gradual destruction has produced these bodies was not discovered until 1866. The August meteors have been observed for more than six centuries, but the comet whose disintegration has furnished the material for this vast stream remains intact, and was not discovered until 1862.

The accepted *comet-meteor* theory does not explain clearly the visibility of comets or the changes that occur in the apparent brightness and in the density of the nucleus as these bodies approach and recede from the sun; neither does it explain in a satisfactory manner the position of the comets in their attendant meteor streams. If comets are composed of solid matter or of discrete solid particles, it would seem quite proper to ask why they become visible at such immense distances from the earth and the sun.

The perihelion distance of 26 per cent. of the comets with known orbits is equal to or greater than the mean distance of the earth from the sun.

Many comets when first seen are much farther from the sun than is the earth at aphelion, and the spectroscope only gives the information that the light is derived from a gas or vapor. From our constant experience with solid masses of stone and iron on the surface of the earth and under the unobstructed influence of the sun, it is impossible to see how the sun's heat alone can produce gas or vapor from such bodies at the observed distances.

As the comet approaches the sun the faint diffused mass of the body begins to contract, and a point in the mass, generally nearer the sun than the center, becomes brighter and denser, frequently, as it rapidly nears the sun, changing its form and brightness in a marked manner from day to day.

It is not improbable that the solid constituents of meteorites would be vaporized if they passed as near the sun as did Comet II, 1882; but it is not probable that this change does occur at distances greater than the radius of the earth's orbit, if it is effected simply by the action of the sun.

If the visibility is caused by the assumed enormous change of temperature experienced by the solid portion of the comet in passing from outer space to the locus of visibility in the solar system, then the entire mass of the comet should be vaporized and solid meteoric bodies would cease to exist.

If, on the other hand, this visibility is brought about by the effect of this change of temperature on the occluded gases stored up in the solid portions of the comet, then during the long period in which these masses are subjected to the solar action these gases would all be expelled and dissipated and none would be found in those meteorites which finally find their way to the surface of the earth and into the chemist's laboratory.

The meteors of the shower of November 27 are scattered along the orbit of that stream for at least 500 millions of miles. If this elongation of the meteor stream is formed, as is highly probable, by the difference in velocity between those meteors on that portion nearest the sun and those on the outside of the mass, then, if the comet is the meteor-producing body, the same action would tend to break it up and destroy it early in its existence as a solar satellite.

If the existence of the comet as a member of the solar system antedates the meteor stream, it is difficult to see how the comet could have remained intact long enough to have been observed, in the presence of forces that for thousands of years have been transforming the figure of the original mass and stretching it out into a stream whose length is measured by hundreds of millions of miles. It is not improbable that comets of large dimensions are destroyed by the action of such forces; but that a body of that character should miraculously survive its own destruction and be found existing in ordinary cometary form in the midst of its own ruins is a proposition that makes large demands on the imagination.

If the brightness of comets is caused by the vaporization of iron or stony matter, it must be produced by collisions between the masses at such velocities that a high temperature is developed, producing an incandescent vapor yielding a distinctive spectrum. It seems difficult to explain how such relative velocities can arise among the individual members of the same stream moving in a common orbit. It is more than probable that the light of a star passing from the extremely low temperature of space through the supposed high temperature of the comet's nucleus, and again into the temperature of space, would suffer so much apparent change of position that it would compel recognition. It is claimed, however, that the individual masses of meteoric matter which form the nucleus are so far separated that the light of a star can pass through the aggregated mass without material change of direction.

But if the masses are vaporized by collisions, then there must be absolute contact, which would to a great extent obstruct the passage of stellar light and would be certain to produce refraction.

Lockyer's Theories.—before leaving the consideration of these points I venture to call attention for a moment to a recent theory which has been set forth with considerable elaboration of detail.

In a paper entitled " Researches on the Spectra of Meteorites," presented to the Royal Society on October 4, 1887,
and in the Bakerian lecture on April 12, 1888, followed by
an appendix to the same on January 10, 1889, Mr. Lockyer
presented in detail his laboratory experiments, combined
with the more or less accurate observations of other astrono-
mers and physicists, which led him to certain definite conclusions in regard to the relations of comets and meteors.

The author's conclusions and theories can be most suc-
cinctly presented in the following citations from the papers
mentioned :

"The existing distinction between stars, comets, and nebulæ
rests on no physical basis."

"All self-luminous bodies in the celestial spaces are composed of meteorites or masses of meteoric vapor produced by
heat brought about by condensation of meteor swarms due
to gravity.".

"Meteorites are formed by the condensation of vapors
thrown off by collisions. The small particles increase by
fusion brought about again by collisions, and this increase
may go on until the meteorites may be large enough to be
smashed by collisions when the heat of impact is not sufficient to produce volatilization of the whole mass."

"Beginning with meteorites of average composition, the
extreme forms, iron and stony, would in time be produced
as the result of collisions."

"The spectra of all such bodies depend upon the heat of
meteorites produced by collisions and the average space between the meteorites in the swarm, or, in the case of consoli-
dated swarms, upon the time which has elapsed since complete vaporization."

"The temperature of vapors produced by collisions in
nebulæ, stars without C and F, but with other bright lines,
and in comets away from perihelion is about that of the
Bunsen burner."

"The temperature of the vapors produced by collisions
in α Orionis and similar stars is about that of the Bessemer
flame."

"The brilliancy of the aggregated masses depends upon the number of the meteorites and not upon the intensity of the light."

"The bright flutings of carbon in the spectra of some 'stars,' taken in conjunction with their absorption phenomena, indicate that widely separated meteorites at a low temperature are involved."

"New stars are produced by the clash of meteor-swarms, the bright lines seen being low temperature lines of those elements in meteorites the spectra of which are most brilliant at a low stage of heat."

"A comet is a swarm of meteors in company. Such a swarm finally makes a continuous orbit by virtue of arrested velocities. Impacts will break up large stones and will produce new vapors, which will condense into small meteoroids."

"When the meteorites are *strongly* heated in a glow-tube the whole tube, when the electric current is passing, gives us the spectrum of carbon. When a meteor-swarm approaches the sun the whole region of space occupied by the meteorites * * * gives us the same spectrum."

"The first stage in the spectrum of a comet is that in which there is only the radiation of the magnesium. The next is that in which Mg. 500 is replaced wholly or partially by the spectrum of cool carbon. Mg. is then added and cool carbon is replaced by hot carbon. The radiation of manganese 558 and sometimes lead 546 is then added. Absorption phenomena next appears, manganese 558 and lead 546 being indicated by thin masking effect upon the citron band of carbon. The absorption band of iron is also sometimes present at this stage. At this stage also the group of carbon flutings, which I have called carbon B, probably also makes its appearance. As the temperature increases still further, magnesium is represented by b, and lines of iron appear. This takes place when the comet is at or near perihelion."

"The observations on meteorites recorded in the Bakerian

Lecture and the discussion of cometary observations contained
in this Appendix show that the vapors which are given out
by the meteorites as the sun is approached are in an approximate order: slight hydrogen, slight carbon compounds,
magnesium, sodium, manganese, lead, and iron.   Now, of
these the hydrogen and carbon compounds are alone permanent gases, and the idea is that they have been occluded
as such by the meteorites."

" The aurora being a low temperature phenomenon, we
should expect to find in its spectrum lines and remnants of
flutings seen in the spectra of meteorites at low temperatures.
The characteristic line of the aurora is the remnant of the
brightest manganese fluting at 558."

" The spectrum of the nebulæ, except in some cases, is
associated with a certain amount of continuous spectrum,
and meteorites glowing at a low temperature would be competent to give the continuous spectrum with its highest intensity in the yellow part of the spectrum."

" Only seven lines in all have been recorded up to the
present in the spectra of nebulæ, three of which coincide
with lines in the spectrum of hydrogen and three correspond to lines in magnesium.   The magnesium lines represented are the ultra-violet low-temperature line at 373, the
line at 470, and the remnant of the magnesium fluting at
500, the brightest part of the spectrum at the temperature
of the Bunsen burner.   The hydrogen lines are h, F, and
Hγ. (434).   Sometimes the 500 line is seen alone, but it is
generally associated with F and a line at 495.   The remain-
ing lines do not all appear in one nebulæ, but are associated
one by one with the other three lines."

" When a tube is used in experiments to determine the
spectrum of meteoric dust at the lowest temperature we find
that the dust in many cases gives a spectrum containing the
magnesium fluting at 500, which is characteristic of the
nebulæ and is often seen alone in them.   If the difference
between nebulæ and comets is merely of cosmographical
position, one being out of the solar system and one being in

it; and, further, if the conditions as regards rest are the same, the spectrum should be the same, and we ought to find this line in the spectrum of comets when the swarm most approaches the undisturbed nebulous condition, the number of collisions being at or near a minimum—*i. e.*, when the comet is near aphelion the fluting should be visible alone."

After citing the results of the spectroscopic observations of several comets, the author remarks: "This spectroscopic evidence is of the strongest, but it does not stand alone. Comets at aphelion present the telescopic appearance, for the most part, of globular nebulæ."

The comprehensive theory set forth in the quotations just cited assumes that the auroræ, nebulæ, meteorites, comets, and most of the stars all have a common origin, and that all the multifarious telescopic and spectroscopic phenomena exhibited by these bodies are due to the varying velocities of the collisions between the meteoric particles and masses of which in some form all these bodies are composed.

We are told that meteorites at a low temperature present in the spectum a certain line, at 558, due to manganese, and also that this line appears in the nebulæ, the aurora, and in comets at considerable distances from perihelion. Hence the identity of all these bodies is inferred and the foundation of the theory is laid.

Meteorites are subjected to laboratory experiments in tubes in which the temperature is gradually raised to a high degree and the varying spectra is noted. Spectroscopic observations of nebulæ, comets, and stars are then compiled and classified, until the several groups are so arranged that they present nearly the same sequence of spectra that have been derived from meteoric matter at increasing temperatures in the experiments.

The theory is then extended and we are given to understand that when, in the case of nebulæ and stars greater activity of collisions occur, or when a comet approaches the sun, the same phenomena appear and in the same order.

The identity of these bodies is then supposed to be complete and the theory established.

This theory of collisions rests upon a remarkable congeries of experiments, observations, and assumptions. Many of the observations and many of the laboratory experiments, which were made by the author, as well as much of the data quoted throughout his papers are entitled to the highest merit. But, considering much of the data and many of the statements in his conclusions, and especially the extraordinary assertion that "comets at aphelion present the telescopic appearance for the most part of globular nebulæ," it is not remarkable to find the author's data, as well as his deductions, vigorously attached by able physicists.

Huggins.— After a careful study of the spectrum of the aurora Mr. *Huggins*[1] remarks: "After consideration, I think that I ought to point out that Mr. Lockyer's recent statement that 'the characteristic line of the aurora is the remnant of the brightest manganese fluting at 558' is clearly inadmissible, considering the evidence we have of the position of this line."

After a very thorough study of the spectra of the nebulæ, Mr. *Huggins*[2] writes: " As, therefore, there seems to be little doubt that the 'remnant of the fluting at 500' is not coincident with the brightest nebular line, and the next most characteristic group of this spectrum, the triplet at 3720, 3724, and 3730, according to *Liveing* and *Dewar*, does not appear to be present in the photographs, we may conclude that the remarkable spectrum of the gaseous nebulæ has not been produced by burning magnesium."

Professor Liveing[3] says in regard to the line denoted by Lockyer as 470: " I have never seen the line at λ 4703 in the spectrum of the magnesium flame. As it is a conspicuous line in the arc and spark, we looked for it in the flame, but did not find it."

[1] Proc. Roy. Soc., XLV, 435.   [2] Proc. Roy. Soc., XLVI, 55.
[3] Proc. Roy. Soc., XLVI, 56.

If the testimony of Huggins and of Liveing and Dewar represents the observed phenomena, and their observations have not yet been disproved, then most of the broad theories of Lockyer, which assume a common origin and structure for auroræ, nebulæ, comets, and stars, lacks a basis of observed facts, resting wholly, so far as the aurora and nebulæ are concerned, on approximate coincidences in the spectra, while the assumed telescopic appearance of comets at aphelion is a creation of the imagination.

CONCLUSIONS.

Attention has been called to these various theories relating to comets and meteors simply with a view to emphasizing the fact that none of the systems, whether simple or complex, seems to explain all the observed phenomena.

As a scientific explanation, the direct and simple is always preferable to the indirect and involved method, and this safe precept should be the guide in all investigations of the apparent physical connection between comets and meteors.

It seems to me that the true theory of the origin and the relations of comets and meteors is yet to be discovered.

When asked to give my own theory of these bodies I can only reply that I have none. At the same time I see less objection to the following hypotheses than to any of those now doing duty as theories:

Meteors and meteorites are solid iron or stony bodies and, whatever their origin, are now members of the solar system. Comets are composed chiefly of gaseous matter, and originate outside of the solar system. Some of these bodies on entering the sphere of solar attraction are so far drawn away from their original orbits by the masses of the sun's outer satellites that they become permanent members of the solar system. Of these, at least *four* have become entangled in the immense aggregations known as meteor streams and have adopted the orbits of their captors. The meteors still remain meteors, however, and the comets retain their former identity and peculiar structure.

Observations of Meteors.

Most of the observers of sporadic meteors and meteorites have been either amateurs or persons entirely deficient in that special training which is so essential in a trustworthy observer of unexpected phenomena. Fortunately, however, most of the important phenomena have been noted by intelligent and skilled observers, whose zeal and care have left little to be desired.

It would be impracticable to mention even the names of all the successful observers, but any sketch of the progress of meteoric astronomy in this country would be notably deficient if some of the prominent names were omitted.

The remarkable meteor shower of November 13, 1833, attracted the attention of many careful observers and zealous students along our Atlantic coast, and for several years the subject was carefully investigated by Prof. Dennison Olmstead and Prof. A. C. Twining, who were the pioneers in the study of this science in the United States.

From 1838 when E. C. Herrick began his work he labored with untiring industry as an observer and a compiler of observations and other data until his death, in 1862, and no one in this country did so much as he in promoting the observation and investigation of the August meteors.

Mr. Herrick also gave considerable attention to the study and observation of the November meteors, but this stream was made a special study by Prof. H. A. Newton, with the best results.

Professor Newton's observations of the November meteors began in 1860 and have been continued to the present time, while his investigations of the motions and character of this stream place him undeniably at the head of American workers in this branch of Astronomy.

Much work of the highest value was done by Prof. C. U. Shepard and by Prof. J. Lawrence Smith in the chemical examination of all classes of meteorites, and excellent investigations of a similar character have been carried out by other eminent chemists in the country.

The zeal and industry of Professor Shepard was shown in his extensive collection of meteoric specimens, which at the time of his death was the largest in America.

The attempt to bring together all the published observations in this country in one systematic collection is a task beset with grave difficulties.

The reports of these observations are scattered through all the scientific journals, the metropolitan and local newspapers, and the proceedings of all grades of learned societies.   Frequently the reports, when found, have but little scientific value from lack of the necessary information.   In many instances much time and space are wasted in describing trivial details which have no interest or value in connection with the true meteoric phenomena, while the really essential data are not mentioned.

It sometimes happens that the only available information in regard to a meteorite is derived from the report of its chemical examination, and there can be found no astronomical data whatever to account for its position; it is simply a portion of the earth's surface, and the *how, when,* and *whence* of its advent remain unanswered.

It has been impossible, sometimes, to find any trustworthy authority for essential data, and it has been necessary frequently to interpret freely where the observer or writer has given but a slight clue to his meaning.

In nearly all cases marginal references are made to the original papers in order to facilitate further examination, if desired.

THE CATALOGUES.

The catalogues of Sporadic Meteors, Meteoric Showers, Observed Meteorites, and Discovered Meteorites are supposed to contain all observations, accompanied with the necessary data that have been found in the various publications to which the author has had access.   It is not assumed, however, that these lists contain all the good observations that have been made in this country; in fact, it is quite certain that they do not, and one of the principal aims of this paper will be attained if this fact attracts sufficient attention to bring to light the missing or the unpublished observations.

In all the catalogues the *day* of the observation is the *astronomical* day. It was manifestly impracticable to give every reference to each object in the five catalogues, and only the most important ones have been retained.

Occasionally references are only given to the first page of a paper when it contains several observations of the same phenomenon.

In the reference notes at the bottom of the page the principal abbreviated notation may be explained as follows:

A. J. S., XXV, 306, refers to the American Journal of Science, Vol. XXV, *second* series, page 306.

Trans. A. P. S. refers to the American Philosophical Society.

Proc. A. P. S. refers to the American Philosophical Society.

Proc. A. A. A. S. refers to the proceedings of the American Association for the Advancement of Science.

CATALOGUES

I.–V.

Number.	Date.				Locality.	Iron or stone.
	Year.	Month.	Day.	Hour. Min.		
1	1781(?)	....	......	......	Portage Bay, Chilcot Inlet, Alaska	I.
2	1807	December	13	18.5	Weston, Conn	S
3	1810	January	30		Caswell, N. C.	S
4	1823	August	7		Nobleborough, Me.	S
5	1825	February	19		Nanjemoy, Md.	S
6	1827	May	9		Sumner Co., Tenn.	S
7	1828	June	4		Richmond, Va.	S
8	1829	May	8	3	Forsyth Co., Ga.	S
9	1829	August	15		Deal, N. J.	S
10	1835	July	30	2	Charlotte, Dickson Co., Tenn.	I
11	1837	May	5	3	East Bridgewater, Mass.	S
12	1839	February	13	3	Little Piney, Pulaski Co., Mo.	S
13	1840	October			Concord, N. H.	S
14	1843	March	25		Bishopsville, S. C.	S
15	1844	January			Argentine Republic	I
16	1846	August	14	3	Cape Girardeau, Mo.	S
17	1847	February	25	3	Marion, Linn Co., Iowa.	S
18	1848	May	19	16	Castine, Me.	S
19	1849	October	31	3	Charlotte, Cabarrus Co., N. C.	S
20	1855	August	5	3	Lincoln Co., Tenn.	S
21	1857	April	1		Costa Rica, Central America.	S
22	1857				Independence Co., Iowa.	S
23	1859	March	28	4	Harrison Co., Ind.	S
24	1859	July	4		Crawford Co., Ark.	S
25	1859	August	11		Bethlehem, N. Y.	S
26	1860	May	1	1	New Concord, Ohio.	S
27	1865	March	24	21	Vernon Co., Wis.	S
28	1868	November	27	5	Danville, Ala.	S
29	1868	December	5	3	Frankfort, Ala.	S
30	1869	October	5	23	Stewart Co., Ga.	S
31	1871	May	20	20	Searsmont, Me.	S
32	1874	May	14	2.5	Nash Co., N. C.	S
33	1875	February	12	10.5	Iowa Co., Iowa.	S
34	1876	June	24	21	Kansas City, Mo.	I
35	1876	December	21	8.75	Rochester, Fulton Co., Ind.	S
36	1877	January	2		Warrenton, Warren Co., Mo.	S
37	1877	January	23		Cynthiana, Harrison Co., Ky.	S
38	1879	May	10	5	Estherville, Emmet Co., Iowa.	S. I
39	1879	August			Fomatlan, Jalisco, Mexico.	I
40	1883				Calderilla, Chili	I
41	1885	November	27		Mazapil, Mexico.	I
42	1886	March	27	3	Johnson Co., Ark.	I
43	1887	January	21	2	De Cewsville, Haldimand Co., Ontario.	M
44	1890	May	2	5 15	Winnebago Co., Iowa.	S

1. Cat. State Mining Bureau of Cal., 1888, No. 2925.
2. { Trans. Amer. Phil. Soc. VI$_1$, 323. ; A. J. S. XXXVII$_1$, 130; VI$_2$, 410.
3. A. J. S. II$_2$, 392.
4. A. J. S. VII$_1$, 170; IX$_1$, 400.
5. A. J. S. IX$_1$, 351; X$_1$, 131; VI$_2$, 406.
6. A. J. S. XVII$_1$, 336; XVIII$_1$, 200, 378.
7. A. J. S. XV$_1$, 195; XVI$_1$, 191.
8. A. J. S. XVIII$_1$, 388.
9. Proc. A. A. S., 1851, Vol. **II, 188.**
10. A. J. S. XLIX$_1$, 336.
11. A. J. S. XXXII$_1$, 395.
12. A. J. S. XXXVII$_1$, 385; **XXXIX$_1$, 254.**
13. A. J. S. IV$_2$, 353; VI$_2$, 416.
14. A. J. S. II$_2$, 392; VI$_2$, 411.
15. Proc. Lit. and Phil. Soc. Liverpool, **VII, 1853.**
16. A. J. S. XXXII$_2$, 229.
17. A. J. S. IV$_2$, 288, 429.
18. A. J. S. VI$_2$, 251, 406.
19. A. J. S. IX$_2$, 143; X$_2$, 127.
20. A. J. S. XXIV$_2$, 134; **XXXI$_2$, 264.**
21. Buchner, 93.
22. A. J. S. XXX$_2$, 208.

Observed Meteorites.

Number.	Weight.	Authority.	Remarks.
1	88 lbs.	....	Seen to fall by the father of one of the oldest Indians.
2	300 "	Nathan Wheeler.................	Observed by many persons.
3	3 "	Madison.	
4	4 "	A. Dinsmoor.	
5	16.5 "	W. D. Harrison.	
6	11 "		
7	4 "		
8	36 "	Elias Beall.	
9	....		Weight "rather more than half an ounce."
10	9 lbs.	G. Troost.....................	J. L. Smith gives the date as August 1.
11	0.5 lb.		
12	50 lbs.	Mr. Harrison.	
13	....		Weight 370.5 grains.
14	13 lbs.	C. U. Shepard.	
15	....	H. E. Symonds...............	Fall witnessed by 1,400 soldiers. About a cubic yard of the mass remained above the surface of the ground.
16	4.5 lbs. {	E. S. Dana.	
		S. L. Penfield.	
17	130 lbs.	D. C. Rogers.	
18	0.1 lb.	Giles Gardner.	
19	19.5 lbs.	H. Bost.	
20	3.9 "	James B. Dooley.	
21			
22	....	C. U. Shepard..................	Fell in the "summer" of 1857.
23	3.7 lbs.		Several observers.
24	....	Mr. Scott.	
25	....	C. U. Shepard	"Smaller than a pigeon's egg."
26	469.2 lbs.		
27	....	J. L. Smith.	
28	4.5 lbs.	W. Brown.	
29	1.7 "	Jas. W. Hooper.	
30	0.8 lb.	Mrs. Buck.	
31	12 lbs.		
32			
33	500 lbs.	....	Many observers.
34			
35	0.8 lb.	A. J. Morris.	
36	100 lbs.		
37	15 "	....	"Fell in the afternoon."
38	750 "		
39	....	C. U. Shepard	Several pieces; the largest weighed about 2 lbs.
40	"Small."	Ward and Howell.............	Not yet described.
41	8.7 lbs.	W. E. Hidden.	
42	107.5 "	G. F. Kunz.	
43	0.75 lb.	E. E. Howell.	
44	181 lbs.	G. F. Kunz..................	Probably more fragments to be discovered.

23. A. J. S. XXVIII₂, 409.
24. Owens' 2d Geolog. Reconnaissance of Arkansas, 408.
25. A. J. S. XXX₂, 396.
26. { A. J. S. XXX₂, 103, 207, 296 ;
    { A. J. S. XXXI₂, 87; XXXII₂, 30.
27. A. J. S. XII₂, 207.
28. A. J. S. XLIX₂, 90.
29. A. J. S. XLVIII₂, 240.
30. A. J. S. I₃, 335, 339.
31. A. J. S. II₃, 133, 200.
32. A. J. S. X₃, 147.
33. A. J. S. IX₃, 407, 459 ; X₃, 44, 357.
34. A. J. S. XII₃, 316.
35. A. J. S. XIII₃, 207, 243.
36. A. J. S. XIII₃, 243; XIV₃, 219.
37. A. J. S. XIII₃, 243; XIV₃, 219.
38. A. J. S. XVII₃, 77, 186; XIX₃, 459, 495; XX₃, 136.
39. A. J. S. XXX₃, 105.
40.
41. A. J. S. XXXIII₃, 221.
42. A. J. S. XXXIII₃, 494, 500.
43. Science, N. Y., March 7, 1890, 167.
44. Science, N. Y., May 16, 1890, 304.

CATALOGUE II.

Number.	Date.		Locality.	Iron or stone.
	Year.	Month and day.		
1	1735		Arizona	I.
2	1784		Bahia, Brazil	I.
3	1792		Zacatecas, Mexico	I.
4	1808		Red River, Texas	I.
5*	1810		Santa Rosa, New Granada	I.
6	1811		Durango, Mexico	I.
7	1818		Lockport, N. Y.	I.
8	1819		Burlington, N. Y.	I.
9	1820		Guilford Co., N. C.	I.
10	1822		Randolph Co., N. C.	I.
11	1826-7		Waterloo, Seneca Co., N. Y.	I.
12	1828		Bedford Co., Pa.	I.
13	1832		Walker Co., Ala.	I.
14	1834		Scriba, Oswego Co., N. Y.	I.
15	1834		Claiborne, Clark Co., Ala.	I.
16	1835		Buncombe Co., N. C.	I.
17	1836		Brazos, Texas	I.
18	1839	March	Putnam Co., Ga.	I.
19	1839		Buncombe Co., N. C.	I.
20	1840	February 26	Chili	I.
21	1840		Cocke Co., Tenn.	I.
22	1840		Smithland, Livingston Co., Ky.	I.
23	1841	February	Lexington Co., S. C.	I.
24	1842		Grayson Co., Va.	I.
25	1842		Roanoke Co., Va.	I.
26	1842		Carthage, Tenn.	I.
27	1842		Green Co., Tenn.	I.
28	1845		DeKalb Co., Tenn.	I.
29	1845		Otsego Co., N. Y.	I.
30	1846		Franconia, N. H.	I.
31	1846		Jackson Co., Tenn.	I.
32	1847		Chesterville, S. C.	I.
33	1847-8		Murfreesborough, Tenn.	I.
34	1849		Pittsburgh, Pa.	I.
35	1850		Allegheny Co., Pa.	I.
36	1850		Seneca River, N. Y.	I.
37	1850		Salt River, Ky.	I.
38	18 0		Botetourt Co., Va.	I.
39	1853		Jefferson Co., Tenn.	I.
40	1853		Union Co., Ga.	I.
41	1853	July	Campbell Co., Tenn.	I.
42	1853	August	Tazewell, Claiborne Co., Tenn.	I.
43	1854		Madoc, Ontario	I.
44	1854		Haywood Co., N. C.	I.
45	1855		Coahuila, Mexico	I.
46	1856		Nelson Co., Ky.	I.
47	1856		Nebraska	I.
48	1856		Madison Co., N. C.	I.
49	1856		Forsyth, Taney Co., Mo.	I.
50	1856		Marshall Co., Ky.	I.
51	1856		Denton Co., Texas	I.
52	1856		Oktibbeha, Miss.	I.
53	1857		Laurens Co., S. C.	I.
54	1858		Washington Co., Wis.	I.

1. { Smithsonian Report 1863, 55, 85.
A. J. S. XVIII₂, 369; XIX₂, 161, 162.
2. { A. J. S. XV₂, 12; XXXVI₂, 158.
Sci. Am. Supp. Oct. 19, 1889.
3. A. J. S. XV₂, 11.
4. A. J. S. VIII₁, 218; XVI₁, 217; XXVII₁, 382.
5. A. J. S. XV₂, 11.
6. A. J. S. XV₂, 19.
7. A. J. S. XLVIII₁, 388; II₂, 374, 391.
8. A. J. S. XLVI₁, 401; II₂, 391; XV₂, 20.
9. A. J. S. XVII₁, 140; XL₁, 369.
10. A. J. S. XVII₁, 140; II₂, 391; XV₂, 21.
11. A. J. S. XI₂, 39; XXXIV₂, 298.
12. A. J. S. II₂, 391; XV₂, 21.
13. A. J. S. XLIX₁, 344; II₂, 391; XV₂, 21.
14. A. J. S. XL₁, 366; II₂, 390; IV₂, 75.

15. A. J. S. XXXIV₁, 332; XLVIII₁, 145.
16. A. J. S. XXXVI₁, 81; XLIII₁, 359.
17. A. J. S. XXXI₁, 459.
18. A. J. S. XVII₂, 331.
19. A. J. S. IV₂, 82.
20. Lon., Ed. and Dublin Phil. Mag. X₄, 12.
21. A. J. S. XXXVIII₂, 280; XLIII₁, 354.
22. A. J. S. II₂, 357; XV₂, 21.
23. { A. J. S. X₂, 128; XV₂, 5, 16.
Proc. A. A. A. S. 1850, Vol. I, 152; 18 Vol. II, 189.
24. A. J. S. XLIII₁, 169; II₂, 392.
25. A. J. S. XLIII₁, 169; II₂, 392.
26. A. J. S. II₂, 356; XV₂, 20.
27. A. J. S. XLIX₁, 342; II₂, 391.

Discovered Meteorites.

Number	Weight.	Authority.	Remarks.
1	1,400 lbs.	Joseph Henry......	The "Ainsa" or "Tucson" meteorite.
2	11,819 "	....	The "Bendego" meteorite.
3	2,000 "		
4	1,635 "		
5	1,700 "		
6	....		Weight 30,000 or 40,000 lbs.
7	36 lbs.	B. Silliman.	
8	150 "	B. Silliman.	
9	28 "	C. U. Shepard.	
10	2 "	C. U. Shepard.	
11	0.1 lb.	C. U. Shepard......	Rammelsburg thought this was not a meteorite.
12	....	C. U. Shepard......	"A few ounces." Dr. Genth thought it was Spiegeleisen.
13	165 lbs.	G. Troost.	
14	8 "	C. U. Shepard.	
15	40 "	C. T. Jackson.	
16	30 "	C. U. Shepard......	One date given is 1845.
17	324 "	C. U. Shepard.	
18	72 "	J. E. Willett.	
19	1.3 "	C. U. Shepard.	
20	17 "	R. P. Greg..........	Found in the desert of Tarapaca.
21	2,000 "	G. Troost.	
22	9.9 "	G. Troost.	
23	117 "	C. U. Shepard......	Found on "Ruff's Mountain."
24	....	J. B. Rogers	"Original mass of many pounds' weight."
25	....	W. B. Rogers......	No weight given.
26	280 lbs.	G. Troost.	
27	29 "	G. Troost..........	Found near Babb's mill.
28	36 "	G. Troost.	
29	....	C. U. Shepard......	Weighed 276 grains.
30	....	Robert Gilmore....	Weighed "15 or 20 pounds."
31	....	G. Troost	Weighed 15 oz.
32	36 lbs.	C. U. Shepard.	
33	19 "	G. Troost.	
34	0.6 lb.	F. A. Genth.	
35	292 lbs.	B. Silliman.	
36	9 "		Date somewhat doubtful.
37	8 "	B. Silliman.	
38	....	C. U. Shepard......	Original mass lost.
39	2.5 lbs.		
40	15 "		
41	0.3 lb.		
42	60 lbs.	C. U. Shepard......	Also described by J. L. Smith.
43	370 "	T. Sterry Hunt.	
44	....	C. U. Shepard......	Weight about ½ oz.
45	252 lbs.	J. L. Smith..........	The "Couch" meteorite.
46	0.2 lb.	G. J. Brush.	
47	35 lbs.		
48	0.1 lb.	G. J. Brush.	
49	197 lbs.	C. U. Shepard.	
50	15 "	C. U. Shepard.	
51	....	C. U. Shepard	Weighed 66 grains.
52	0.3 lbs.	W. J. Taylor..........	Found in an Indian mound.
53	4.7 "	W. E. Hidden.	
54	85.8 "	F. Breundecke.	

28. A. J. S. XLIX₁, 341; II₂, 391.
29. A. J. S. II₂, 391; XV₂, 16.
30. A. J. S. II₂, 392; IV₂, 87.
31. A. J. S. II₂, 357; XV₂, 21.
32. A. J. S. VII₂, 449; XV₂, 21.
33. A. J. S. V₂, 351; XV₂, 21.
34. A. J. S. XV₂, 22; XII₂, 72.
35. A. J. S. XV₂, 7; Proc. A. A. A. S. 1850, Vol. II, 37.
36. A. J. S. XIV₂, 439; XV₂, 363.
37. A. J. S. XV₂, 22; Proc. A. A. A. S. 1850, 36.
38. A. J. S. XLII₂, 250.
39. A. J. S. XVII₂, 329.
40. A. J. S. XVII₂, 328.
41. A. J. S. XIX₂, 153.

42. A. J. S. XVII₂, 131, 325; XIX₂, 153.
43. A. J. S. XIX₂, 417.
44. A. J. S. XVII₂, 327.
45. { A. J. S. XIX₂, 160.
     { Smithsonian Report, 1863, 56.
46. A. J. S. XXX₂, 240; XXXI₂, 459.
47. A. J. S. XXX₂, 204; XXXII₂, 146.
48. A. J. S. XXX₂, 240; XXXI₂, 459.
49. A. J. S. XXX₂, 205; XXXIV₂, 467.
50. A. J. S. XXX₂, 240; XXXI₂, 459.
51. { A. J. S. XXXI₂, 459.
     { Trans. St. Louis Acad. Sci. I, 623.
52. A. J. S. XXIV₂, 293.
53. A. J. S. XXXI₂, 463.
54. Smithsonian Report 1869, 417.

CATALOGUE II.—Dis

Number.	Date		Locality.	Iron or stone.
	Year.	Month and day.		
55	1858-9		Augusta Co., Va	I.
56	1859		Rogue River Mountains, Ore	I.
57	1860		Mountains of East Tennessee	I.
58	1860		Franklin Co., Ky	I.
59	1860	October	La Grange, Oldham Co., Ky	I.
60	1860	December	Coopertown, Tenn	I.
61	1860		Newton Co., Ark	I.
62	1861		Rensselaer Co., N. Y	S.
63	1861	February 18	Colorado	I.
64	1863		Tucson, Arizona	I.
65	1863	June 9	Dakota	I.
66	1866		Bear Creek, Colo	I.
67	1866		Frankfort, Ky	I.
68	1867		Allen County, Ky	I.
69	1868	April	Losttown, Cherokee Co., Ga	I.
70	1868		"Southeastern Missouri"	I.
71	1868		Auburn, Macon Co., Ala	I.
72	1869		Utah	S.
73	1869		El Dorado Co., Cal	I.
74	1869		Trenton, Wis	I.
75	1870		Howard Co., Ind	I.
76	1873	August	Madison Co., N. C	I.
77	1873		Cleburne Co., Ala	I.
78	1874		Waconda, Kan	S.
79	1875		San Francisco, Brazil	I.
80	1877		Whitfield Co., Ga	I.
81	1878		Fayette Co., Texas	S.
82	1879		Whitfield Co., Ga	I.
83	1879	July 19	Davidson Co., N. C	I.
84	1879		Ivanpah, Cal	I.
85	1880		Eagle Station, Carroll Co., Ky	I.
86	1880		Lexington Co., S. C	I.
87	1880		Rutherford Co., N. C	I.
88	1882	June 10	Maverick Co., Texas	I.
89	1882		Burke Co., N. C	I.
90	1883	May 15	Grand Rapids, Mich	I.
91	1883		Little Miami Valley, Ohio	I.
92	1883		Wayne Co., W. Va	I.
93	1884	June	Independence Co., Ark	I.
94	1884		Hammond, St. Croix Co., Wis	I.
95	1884	August	Santa Fé County, New Mexico	I.
96	1884		Chili	I.
97	1885		Catorze, San Luis Potosi, Mex	I.
98	1887	January	Laramie Co., Wyoming	I.
99	1887	March	Claiborne Co., Tenn	I.
100	1887	March	Cumberland Co., Tenn	S.
101	1887	March 27	Chattooga Co., Ga	I.
102	1888	April 30	Welland, Ontario	I.
103	1888		Chili	I.
104	1888		Chili	I. & S
105	1888		Chili	I.
106	1888		Hamilton Co., Texas	I.

55. A. J. S. XV_2, 337.
56. Proc. Boston Soc. Nat. Hist. Vol. 7, 161, 174, 175, 191, 279, 289.
57. { A. J. S., XXXIV_2, 473. / Proc. Acad. Nat. Sci. Phil. 1886, 366.
58. Smith. Report 1868, 343; A. J. S. XLIX, 331.
59. A. J. S. XXXI_2, 151, 265.
60. A. J. S. XXXI_2, 151, 266.
61. A. J. S. XL_2, 213; XXXIV_2, 471.
62. A. J. S. XXXIV_2, 60.
63. A. J. S. XLII_2, 218.
64. { A. J. S. XXXVI_2, 152. / Proc. Cal. Acad. Sci. III, 30.
65. A. J. S. XXXV_2, 259.
66. A. J. S. XLII_2, 250, 286; XLIII_2, 280.

67. A. J. S. XLIX_2, 331.
68. A. J. S. XXXIII_2, 500.
69. A. J. S. XLVI_2, 267; XLVII_2, 294.
70. A. J. S. XLVII_2, 232.
71. A. J. S. XLVII_2, 230.
72. A. J. S. XXXII_2, 225.
73. A. J. S. III_2, 438; VI_2, 18.
74. A. J. S. XLVII_2, 271; III_2, 69.
75. A. J. S. V_2, 155; VII_2, 391.
76. A. J. S. XII_2, 439.
77. A. J. S. XIX_2, 370; XX_2, 74.
78. A. J. S. XI_2, 473; XIII_2, 211.
79. { Comptes Rendus LXXXIII, 917, 918 / LXXXIV, 478, 482, 1065, 1568. / A. J. S. XXIII_2, 232; XXIX_2, 33, 496.

covered Meteorites—Cont'd.

Number.	Weight.	Authority.	Remarks.
55	152 lbs.	J. W. Mallet.	
56	....	John Evans	Mass above ground, 4.5 x 3.5 feet.
57	254 lbs.	F. A. Genth.	
58	0.1 lb.	G. J. Brush.	
59	112 lbs.	....	Found by Wm. Daring.
60	37 "	....	Found by D. Crockett.
61	....	J. L. Smith	Specimen weighed 22½ oz.
62	3.3 lbs.	S. C. H. Bailey.	
63	29 "	....	Found near Central City by Otho Curtice.
64	632 "		
65	10.6 "	Dr. Jackson	Found in the " Dakota Indian country."
66	436 "	J. L. Wilson.	
67	24 "	J. L. Smith.	
68	24.3 "	J. E. Whitfield.	
69	6.6 "		
70	0.8 lb.	C. U. Shepard.	
71	8 lbs.	C. U. Shepard.	
72	1.9 "	E. S. Dana.	
73	85 "	B. Silliman.	
74	143.5 "	J. L. Smith	Six fragments found; the first in 1809.
75	4 "	E. T Cox.	
76	25 "	B. S. Burton.	
77	35.8 "	W. E. Hidden.	
78	116 "	C. U. Shepard.	
79	22,048 "	E. Guignet	In the province of San Catherina.
80	13 "	W. E. Hidden.	
81	321 "	Whitfield and Merrill.	
82	117 "	C. U. Shepard.	
83	2.8 "	W. E. Hidden.	
84	128.2 "	C. U. Shepard.	
85	80 "	G. F. Kunz.	
86	10.5 "	C. U. Shepard.	
87	4.8 "	L. G. Eakins.	
88	97.25 "	W. E. Hidden.	
89	1 lb.	G. F. Kunz.	
90	114 lbs.	J. R. Eastman.	
91	....	G. F. Kunz	Fragments found in mounds by F. W. Putnam; now in the Peabody Museum.
92	26 lbs.	G. F. Kunz	Several fragments.
93	94 "	W. E. Hidden.	
94	53 "	Davenport Fisher.	
95	324.4 "	G. F. Kunz	Several fragments.
96	14.5 "	Ward and Howell.	Found near Puquios. Not yet described.
97	92 "	G. F. Kunz.	
98	25.06 "	G. F. Kunz.	
99	18 "	G. F. Kunz.	
100	94.5 "	J. E. Whitfield.	
101	27 "	G. F. Kunz.	
102	17.5 "	E. E. Howell.	
103	16 "	Ward and Howell.	Thirty leagues east of Taltal. Not yet described.
104	27 "	Ward and Howell.	Thirty-five leagues S. E. of Taltal. Not yet described.
105	....	Ward and Howell.	Estimated at from 6 to 8 lbs.
106	179 "	Ward and Howell.	Found five miles south of Carlton. Not yet described.

80. A. J. S. XIV_2, 246; XXI_2, 286.
81. A. J. S. XXXVI_2, 113.
82. A. J. S. XXVI_2, 336; XXXIV_2, 473.
84. A. J. S. XX_2, 324.
84. A. J. S. XIX_2, 381.
85. A. J. S. XXXIII_2, 228.
86. A. J. S. XXI_2, 117.
87. A. J. S. XXXIX_2, 395.
88. A. J. S. XXXII_2, 394; XXXIII_2, 115.
89. A. J. S. XXXVI_2, 275.
90. A. J. S. XXVIII_2, 299; XXX_2, 312.
91. A. J. S. XXXIII_2, 228.
92. A. J. S. XXXI_2, 145; Proc. A. A. S., 1885, 246.

93. A. J. S. XXXI_2, 460; School of Mines Quarterly, Columbia Coll., VII, No. 2, 168.
94. A. J. S. XXXIV_2, 381.
95. A. J. S. XXX_2, 235; XXXII_2, 311.
96.
97. A. J. S. XXXIII_2, 233.
98. A. J. S. XXXVI_2, 276.
99. A. J. S. XXXIV_2, 475.
100. A. J. S. XXXIV_2, 387, 476.
101. A. J. S. XXXIV_2, 471.
102. Science, N. Y., March 7, 1890, 167.
103.
104.
105.
106.

CATALOGUE III.—Discovere

Number.	Locality.	Iron or stone.	Weight.
1		I.	1 lb.
2	Tucuman, Argentine Republic	I.	30,000 lbs.
3	Crawford Co., Arkansas	S.	1.4 "
4	British America	I.	386 "
5	Canyon City, Trinity Co., Cal.	I.	49 "
6	Los Angeles, Cal.	I.	80 "
7	San Bernardino Co., Cal.	S.	
8	Chihuahua	I.	3,853 lbs.
9	Atacama, Chili	I.	
10	Sierra de Chaco, Chili	S. & I.	
11	Durango, Mexico	I.	72.75 "
12	Oaxaca, Mexico	I.	
13	San Luis Potosi, Mexico	I.	
14	Xiquipilco, Mexico	I.	108.6 "
15	Mexico	I.	192 "
16	Mexico	I.	
17	Mexico	I.	
18	Mexico	I.	2,942 lbs.
19	Mexico	I.	0.5 lb.
20	Mexico	I.	
21	Mexico	I.	5,000 lbs.
22	Bates Co., Missouri	I.	90 "
23	Ironton, Missouri	I.	
24	Rasgata, New Granada	I.	129 lbs.
25	Rockingham Co., N. C.	I.	11 "
26		I.	6 "
27	Green Co., Tenn	I.	290 "
28	Jefferson Co., Tenn	I.	1.4 "
29		S.	6.5 "
30	Augusta Co., Va.	I.	3.5 "
31	Augusta Co., Va.	I.	36 "
32	Augusta Co., Va.	I.	56 "
33	Augusta Co., Va.	I.	2.2 "
34	Kiowa Co., Kansas	I.	
35	Chili	I.	95.5 lbs.
36	Jackson Co., Oregon	I.	2 oz.

1. A. J. S. XXXIV₂, 59.
2. A. J. S. XV₂, 12.
3. Owen's 2d Geological Reconnaissance of Ark., 408.
4. Trans. Roy. Soc. Canada, IV, sect. III, 97.
5. A. J. S., XXIX₂, 489.
6. A. J. S. IV₂, 495.
7. A. J. S. XXXV₂, 490.
8. { A. J. S. XIX₂, 163 ; II₃, 335; III₃, 207.
   { Proc. A. A. A. S. 1871, 266.
9. Buchner, 127.
10. Buchner, 131.
11. A. J. S. XXXVII₂, 439.
12. A. J. S. XV₂, 21.
13. Buchner, 149.
14. A. J. S. XV₂, 20; XXII₂, 374; XXIV₃, 295.
15. A. J. S. XXIX₂, 232.
16. Proc. A. A. A. S. 1871, 269.
17. Proc. A. A. A. S. 1871, 269.

Meteorites without Date.

Number.	Authority.	Remarks.
1	R. B. Riggs....................	Found in Col. Abert's collection of minerals presented to the National Museum.
2		
3		
4	A. P. Coleman	Brought to Coburg, Canada, in 1869.
5	C. U. Shepard.	
6	C. T. Jackson.	
7	G. P. Merrill..............	Put through an ore-crusher before its character was known.
8	J. L. Smith.	
9		
10		
11	J. E. Whitfield.	
12		
13		
14	....	Numerous pieces.
15	N. T. Lupton.	
16	J. L. Smith	The San Gregorio meteorite, 6.5 feet long, 5.5 wide, 4.0 high.
17	J. L. Smith	"The largest yet found in that vicinity."
18	J. L. Smith	The "Butcher" meteorites—six, weighing 290, 430, 438, 550, 580, and 654 lbs.
19	J. L. Smith	Found in a collection of minerals from Mexico.
20	....	Now in National Museum, Washington, D. C. Estimated weight, 2,500 lbs.
21	W. M. Pierson.	
22	G. C. Brodhead.	
23	....	Small specimen.
24		
25	J. L. Smith.	
26	C. U. Shepard..............	Property of the Smithsonian Institution; place of discovery unknown.
27	W. P. Blake.	
28	R. B. Riggs	Riggs considers this a doubtful specimen.
29	L. G. Eakins..............	Probably found in Texas.
30	J. W. Mallet.	
31	J. W. Mallet.	
32	J. W. Mallet.	
33	G. F. Kunz and J. W. Mallet.	
34	F. H. Snow	Found by "cowboys" before that portion of Kansas was settled; fragments weighing in the aggregate more than 1,600 lbs. discovered.
35	Ward and Howell..........	Found about 10 or 12 leagues east of the port of Chañaral; not yet described.
36	Ward and Howell..........	Piece of a mass found by a miner.

18. Proc. A. A. S. 1871, 269.
19. A. J. S. XLV$_2$, 77
20.
21. Smithsonian Report, 1873, 419.
22. A. J. S. X$_2$, 401; XIII$_2$, 213.
23.
24. A. J. S. XV$_2$, 11.
25. A. J. S. XIII$_2$, 213.
26. A. J. S. XXII$_2$, 119.
27. A. J. S. XXXI$_2$, 41.
28. A. J. S. XXXIV$_2$, 60.
29. A. J. S. XXXIX$_2$, 59.
30. A. J. S. II$_3$, 10.
31. A. J. S. II$_3$, 10.
32. A. J. S. II$_3$, 10.
33. A. J. S. XXXIII$_2$, 58.
34. Science N. Y., Vol. XV, No. 379, 290; No. 384, 359
35.
36.

CATALOGUE IV.—

Ref. number.	Date.			Time of Showers.			Time of max. flight.	Whole number counted.	Max. hourly rate.	Radiant point.	No. of observ's.
	Year.	Month.	Day.	Begin-ning.	End.						
				h. m.	h. m.	h. m.					
1	1803	April...	19	13 0	15 0				800		
2	1833	Nov.....	13	9 0	Sunrise	16 0	200,000		"Bend of the Sickle" in Leo	
3	1834	" ...	12	14 30	17 25	17 0	18			1
4	1834	" ...	12				Many.		"In Leo"	
5	1835	" ...	13							
6	1835	" ...	13							
7	1835	" ...	13							
8	1836	" ...	12	14 0	16 0		248			
9	1836	" ...	12	15 0	18 15		253			
10	1836	" ...	12	15 30	16 0	15 40	29			
11	1836	" ...	12				23			1
12	1836	" ...	12				75			
13	1836	" ...	12				500			
14	1836	" ...	12				18			
15	1837	Aug....	9	8 0	15 0				α = 55°, δ = + 60°	1
16	1837	Nov.....	12	13 5	Sunrise	16 30	226		Near μ Leonis	8
17	1837	" ...	12	14 0	"		70		In Leo	1
18	1837	" ...	12	14 0	18 0		45		"	1
19	1837	" ...	12	14 45	16 37		74			12
20	1837	" ...	12	15 15	17 15	16 0	34	17		1
21	1837	" ...	12	15 45	17 0		52		In Leo	1
22	1838	April...	20	10 0	16 0		154		"	2
23	1838	Aug....	8	11 30	12 30		20		α = 55°, δ = + 60°	1
24	1838	" ...	9	9 30	16 20	16 15	54		α = 35°, δ = + 69°	1
25	1838	" ...	9	15 0	15 45		24			1
26	1838	" ...	10	8 55	10 0		36		In Cassiopeæ	3
27	1838	" ...	10	9 0	11 0		48			2
28	1838	" ...	10	9 30	16 0		122		Near ε Cassiopeæ	1
29	1838	" ...	10	14 30	16 0		65		"Sword handle of Perseus"	2
30	1838	Nov....	11	12 0	18 0	15 30	199	53	In Leo	8
31	1838	" ...	12	12 0	17 0	13 30	131	37	"	8
32	1838	" ...	13	13 0	17 20	15 30	233	80	"	8
33	1838	" ...	22	15 20	15 30	14 30	50	25		8
34	1838	Dec....	6	8 0	17 0		113		In Perseus.	5
35	1838	" ...	7	8 0	11 0		210		" "	5
36	1838	" ...	7	10 0	11 0		78		"Zenith"	3
37	1838	" ...	8	7 15	9 0		59			3
38	1838	" ...	11	8 45	10 0		18			1
39	1838	" ...	12	6 0	13 30		28			3
40	1838	" ...	15	6 0	9 15		9			4
41	1839	April...	18	12 0	15 0		58		α = 273°, δ = + 45°	2
42	1839	" ...	18	14 0	16 0		25			2
43	1839	" ...	19	12 15	13 15		19			4
44	1839	" ...	19	14 0	16 0		25		Vicinity of Cassiopeæ	2
45	1839	Aug. ...	4	9 0	11 0		36		Vicinity of Cassiopeæ	2
46	1839	" ...	9	9 7	14 7	13 30	691	194	Sword handle of Perseus	4
47	1839	" ...	9	12 41	15 36		100			1
48	1839	" ...	10	9 0	16 0		187	60		1
49	1839	" ...	10	9 0			76			2
50	1839	" ...	10	9 30	10 0		33		"Algenib"	1

1. A. J. S. XXXVI₁, 358; Va. Gazette and General Advertiser, April 23, 1803; N. H. Gazette, May 31, 1803; Med. Repository, N. Y., Vol. I, 1804.
2. A. J. S. XXV₁, 354, 363.
3. A. J. S. XXVII₁, 355; XXVIII₁, 305.
4. A. J. S. XXVII₁, 339.
5. A. J. S. XXX₁, 375.
6. " " " "
7. A. J. S. XXX₁, 376.
8. A. J. S. XXXI₁, 390.
9. A. J. S. XXXI₁, 388.
10. A. J. S. XXXI₁, 390.
11. A. J. S. XXXI₁, 389.
12. A. J. S. XXXI₁, 391.
13. A. J. S. XXXI₁, 392.
14. A. J. S. XXXII₁, 392.
15. A. J. S. XXXIII₁, 138.
16. A. J. S. XXXIII₁, 379.
17. " " " "
18. " " " "
19. " " " "
20. " " " "
21. " " " "
22. A. J. S. XXXIV₁, 398.
23. A. J. S. XXXV₁, 167.
24. " " " "

Meteor Showers.

Ref. number.	Authority.	Place of observation.	Remarks.
1		Richmond, Va., and Portsmouth, N. H.	"Fell too fast to be counted."
2	Prof. D. Olmstead	New Haven, Conn	Number estimated; seen throughout the Atlantic and Gulf States.
3	Prof. A. D. Bache	Philadelphia, Pa.	
4	Prof. A. C. Twining	West Point, N. Y	Observed on the morning of the 13th.
5		Amenia, N. Y.	"Unusual number;" hourly rate estimated at 40 or 50.
6		Mt. St. Mary's College, Md.	"Unusual number." Observed on the morning of the 14th.
7		Salisbury, N. C.	"Unusual number."
8		New York, N. Y.	Many seen that were not recorded.
9	S. Dunster	Springvale, Me.	
10		New Haven, Conn	Cloudy.
11		Cambridge, Mass.	Observed one hour.
12		Newark, N. J.	Whole number estimated at 400.
13		Randolph and Macon College, Va.	
14	J. L. Russell	Hingham, Mass.	
15	G. C. Schaeffer	New York, N. Y.	Counted between 200 and 300.
16	Prof. D. Olmstead	New Haven, Conn	Majority left trains.
17	G. C. Schaeffer	New York, N. Y.	Many trains.
18	Prof. F. A. P. Barnard.	New York, N. Y.	
19	Prof. E. Loomis	Hudson, Ohio.	
20	E. Fitch	New Haven, Conn.	
21	L. Obermeyer	Mt. St. Mary's College, Md.	
22	Prof. Wright	Knoxville, Tenn.	
23	G. C. Schaeffer	Barren Hill, Pa.	
24	T. R. Dutton	Wilmington Island, Ga.	
25	W. A. Sparks	Society Hill, S. C.	
26	J. D. Dana	Norfolk, Va.	
27	E. C. Herrick	New Haven, Conn.	
28	T. R. Dutton	Wilmington Island, Ga.	
29	C. G. Forshey	Rock Island, Ill.	
30	Prof. J. Lovering	Cambridge, Mass.	Only two observers after 17 hours.
31	Prof. J. Lovering	Cambridge, Mass.	
32	Prof. J. Lovering	Cambridge, Mass.	
33	Prof. J. Lovering	Cambridge, Mass.	
34	E. C. Herrick	New Haven, Conn.	
35	E. C. Herrick	New Haven, Conn.	
36	Prof. A. W. Smith	Middletown, Conn.	
37	E. C. Herrick	New Haven, Conn.	
38	E. C. Herrick	New Haven, Conn.	
39	E. C. Herrick	New Haven, Conn.	
40	E. C. Herrick	New Haven, Conn.	
41	E. C. Herrick	New Haven, Conn.	
42	Prof. E. Loomis	Hudson, Ohio.	
43	E. C. Herrick	New Haven, Conn.	
44	Prof. E. Loomis	Hudson, Ohio.	
45	E. C. Herrick	New Haven, Conn.	
46	E. C. Herrick	New Haven, Conn.	
47	T. R. Dutton	Columbia, Tenn.	
48	L. L. Knox	Middletown, Conn.	
49	Charles Baldwin	New York, N. Y.	
50	Prof. A. W. Smith	Middletown, Conn.	

25. A. J. S. XXXV₁, 167.
26. " " "
27. " " "
28. " " "
29. Trans. Am. Phil. Soc. VII, 296.
30. A. J. S. XXXV₁, 323.
31. " " "
32. " " "
33. " " "
34. A. J. S. XXXV₁, 361; XXXVI₁, 355.
35. A. J. S. XXXV₁, 361.
36. " " "
37. " " "

38. A. J. S. XXXV₁, 361.
39. " " "
40. " " "
41. A. J. S. XXXVI₁, 361.
42. " " "
43. " " "
44. " " "
45. A. J. S. XXXVII₁, 325.
46. A. J. S. XXXVII₁, 325.
47. A. J. S. XXXVIII₁, 203.
48. A. J. S. XXXVI₁, 325.
49. " " "
50. " " "

CATALOGUE IV.—

Ref. number.	Date			Time of Shower.				Time of max. flight.	Whole number counted.	Max. hourly rate.	Radiant point.	No. of observ's.
	Year.	Month.	Day.	Begin- ning.		End.						
				h.	m.	h.	m.	h. m.				
51	1839	Aug. ...	10	10	0	13	0	13 0	491	189	Sword handle of Perseus.....	4
52	1839	" ...	10	11	20	12	20	....	50	....	" "	1
53	1839	" ...	11	9	0	10	0	....	28	....	" "	5
54	1839	" ...	14	10	6	12	0	....	72	44	Bet. Cassiopeæ and Perseus..	3
55	1840	" ...	9	8	0	16	0	....	171	....	Sword handle of Perseus.....	1
56	1840	" ...	9	10	0	15	30	15 0	818	332	" "	4
57	1840	" ...	9	14	15	14	45	....	112	76	α = 30°, δ = + 63° 30'.....	1
58	1840	" ...	10	15	0	15	35	....	35	....	" "	1
59	1841	April...	18	8	30	11	30	....	60	....	α = 196°, δ = − 8°.....	1
60	1841	" ...	19	11	0	13	0	....	20	....		3
61	1841	Aug. ...	9	12	0	15	0	....	40	....		1
62	1841	" ...	10	9	0	10	0	....	60	....		2
63	1842	April...	20	10	20	16	0	15 30	151	55	Corona Borealis	5
64	1842	Aug. ...	8	9	50	16	0	....	90	....		7
65	1842	" ...	9	10	0	12	0	....	133	79		7
66	1842	" ...	10	10	10	11	0	....	89	....	Sword handle of Perseus.....	7
67	1842	Nov.....	13	15	0	16	0	....	46	....		2
68	1844	Aug. ...	9	11	20	15	0	13 30	367	139		4
69	1844	" ...	10	8	0	13	0	....	117	....	β Cassiopeæ.	1
70	1844	" ...	11	9	50	15	0	14 0	622	152		4
71	1844	" ...	12	13	50	15	0	....	46	....	Sword handle of Perseus.....	2
72	1845	" ...	9	10	0	13	0	....	81	....		1
73	1846	" ...	10	12	0	14	6	....	46	....		6
74	1846	" ...	11	9	0	10	0	....	41	....		2
75	1847	" ...	10	10	0	12	0	....	41	....	Head of Perseus.....	1
76	1847	" ...	10	12	0	14	0	....	415	....		4
77	1847	" ...	11	9	15	10	0	....	37	....		4
78	1848	" ...	9	13	0	15	30	15 15	475	281	In Perseus	4
79	1848	" ...	9	13	30	15	30	....	216	....		3
80	1848	" ...	9	15	0	16	45	....	55	....	In Perseus	1
81	1849	April...	19	13	0	14	0	....	54	....		3
82	1849	Aug. ...	10	10	0	12	30	....	260	....		4
83	1850	" ...	9	12	40	15	0	13 30	451	216	In Perseus	3
84	1850	" ...	10	10	0	12	0	....	312	....		3
85	1850	" ...	10	12	0	15	0	....	351	....		2
86	1852	" ...	9	14	0	14	40	....	19	....		1
87	1853	" ...	10	11	45	15	25	14 30	408	119	Sword handle of Perseus.....	3
88	1855	" ...	9	10	25	14	30	13 30	385	110	" "	3
89	1855	" ...	10	10	30	13	0	13 30	290	136	" "	3
90	1855	" ...	11	10	0	11	0	....	37	....	" "	1
91	1855	" ...	11	13	55	15	5	....	58	....	" "	1
92	1855	" ...	12	13	30	14	45	....	36	....	" "	1
93	1856	" ...	10	11	5	14	50	12 30	283	100	" "	3
94	1858	" ...	9	10	45	15	10	14 35	128	....	" "	1
95	1858	" ...	9	12	5	14	5	....	56	....	" "	1
96	1858	" ...	10	8	30	9	0	....	11	....	" "	1
97	1858	" ...	10	13	7	13	52	....	29	....	" "	1
98	1858	" ...	11	14	15	15	15	....	33	....	" "	1
99	1859	July ...	20	10	30	11	30	....	16	....		1
100	1859	Aug. ...	5	11	0	12	0	....	19	....		1
101	1859	" ...	9	11	0	13	0	....	64	....		3
102	1859	" ...	9	13	0	15	30	15 15	304	156	Sword handle of Perseus.....	4
103	1859	" ...	9	14	15	15	30	....	56	....	α = 38° 30', δ = + 57° 15' ..	1
104	1860	" ...	9	10	0	15	10	14 30	588	155	Sword handle of Perseus.....	7

51. A. J. S. XXXVII₁, 325.
52. Trans. Am. Phil. Soc. VII, 268.
53. " " " "
54. A. J. S. XXXVII₁, 325.
55. Trans. Am. Phil. Soc. VII, 270.
56. A. J. S. XXXIX₁, 328.
57. " " " "
58. " " " "
59. A. J. S. XLII₁, 397.
60. " " " "
61. A. J. S. XLI₁, 399.
62. " " " "
63. A. J. S. XLIII₁, 212.
64. A. J. S. XLIII₁, 377.

65. A. J. S. XLIII₁, 377.
66. " " " "
67. A. J. S. XLIV₁, 209.
68. A. J. S. XLVII₁, 316.
69. A. J. S. XLVII₁, 320.
70. A. J. S. XLVIII₁, 316.
71. " " " "
72. A. J. S. I₂, 86.
73. A. J. S. III₂, 125.
74. " " " "
75. Sidereal Messenger, II, 14.
76. A. J. S. VI₂, 278.
77. " " " "
78. A. J. S. VI₂, 279.

Meteor Showers—Cont'd.

Ref. number.	Authority.	Place of observation.	Remarks.
51	E. C. Herrick	New Haven, Conn.	
52	C. G. Forshey	St. Louis, Mo.	
53	C. G. Forshey	Illinois River.	
54	E. C. Herrick	New Haven, Conn.	
55	C. G. Forshey	Philadelphia, Pa.	
56	E. C. Herrick	New Haven, Conn	Moon set at 14h. 0m.
57	G. C. Schaeffer	Jamaica, L. I.	
58	G. C. Schaeffer	Jamaica, L. I.	
59	C. G. Forshey	Vidalia, La.	No trains; paths short.
60	E. C. Herrick	New Haven, Conn.	
61	Dr. J. S. Huntington, U. S. N.	Pensacola, Fla.	
62	Dr. John Locke	Cincinnati, Ohio.	
63	E. C. Herrick	New Haven, Conn	Moon set at 15h. 0m.
64	E. C. Herrick	New Haven, Conn.	Cloudy; actual observing time 1h. 10m.
65	E. C. Herrick	New Haven, Conn	Cloudy.
66	E. C. Herrick	New Haven, Conn	Cloudy.
67	E. C. Herrick	New Haven, Conn	Cloudy after 16h.
68	E. C. Herrick	New Haven, Conn	Partially cloudy.
69	S. R. Williams	Canonsburg, Pa.	
70	E. C. Herrick	New Haven, Conn.	
71	E. C. Herrick	New Haven, Conn.	
72	E. C. Herrick	New Haven, Conn	Cloudy.
73	E. C. Herrick	New Haven, Conn.	
74	E. C. Herrick	New Haven, Conn.	
75	C. G. Forshey	Pass Christian, La	33 conformable with the radiant.
76	W. M. Smith	Manlius, N. Y	Cloudy after 14h.
77	E. C. Herrick	New Haven, Conn	Completely cloudy after 10h.
78	E. C. Herrick	New Haven, Conn.	
79	E. C. Herrick	"On Mt. Carmel."	
80	C. G. Forshey	Mouth of Miss. river.	
81	E. C. Herrick	New Haven, Conn.	
82	S. R. Williams	Canonsburg, Pa.	
83	E. C. Herrick	New Haven, Conn.	
84	E. C. Herrick	New Haven, Conn.	
85	E. C. Herrick	New Haven, Conn.	
86	John Edmunds	New Haven, Conn.	
87	E. C. Herrick	New Haven, Conn.	
88	E. C. Herrick	New Haven, Conn	306 conformable.
89	E. C. Herrick	New Haven, Conn	242 conformable.
90	E. C. Herrick	New Haven, Conn	One-half conformable.
91	E. C. Herrick	New Haven, Conn	45 conformable.
92	E. C. Herrick	New Haven, Conn	20 conformable.
93	E. C. Herrick	New Haven, Conn.	
94	F. Bradley		Observed on railroad between Davenport and Chicago, Ill.
95	Prof. A. C. Twining	Cleveland, Ohio.	
96	Prof. A. C. Twining	Cleveland, Ohio.	
97	Prof. A. C. Twining	Cleveland, Ohio.	
98	Prof. A. C. Twining	Cleveland, Ohio.	
99	F. Bradley	Chicago, Ill	6 conformable to the August radiant
100	F. Bradley	Chicago, Ill	8 conformable to the August radiant.
101	F. Bradley	Chicago, Ill.	
102	F. Bradley	Chicago, Ill	Only a few unconformable meteors.
103	Prof. A. C. Twining	Boston, Mass.	
104	E. C. Herrick	New Haven, Conn.	

79. A. J. S. VI₂, 279.
80. A. J. S. XI₂, 133.
81. A. J. S. VIII₂, 429.
82. "   "   "   "
83. A. J. S. XI₂, 130.
84. "   "   "   "
85. "   "   "   "
86. A. J. S. XIV₂, 430.
87. A. J. S. XVI₂, 288.
88. A. J. S. XX₂, 285.
89. "   "   "   "
90. "   "   "   "
91. A. J. S. XX₂, 285.

92. A. J. S. XX₂, 285.
93. A. J. S. XXII₂, 290.
94. A. J. S. XXVI₂, 435.
95. "   "   "   "
96. "   "   "   "
97. "   "   "   "
98. "   "   "   "
99. A. J. S. XXVIII₂, 446.
100. "   "   "   "
101. "   "   "   "
102. "   "   "   "
103. "   "   "   "
104. A. J. S. XXX₂, 296.

CATALOGUE IV.—

Ref. number	Date			Time of Shower		Time of max. flight.	Whole number counted.	Max. hourly rate.	Radiant point.	No. of observ's.
	Year.	Month.	Day.	Begin- ning.	End.					
				h. m.	h. m.	h. m.				
105	1860	Aug.	9	11 30	13 0	....	57	....	Sword handle of Perseus........	1
106	1860	"	10	11 0	14 0	12 30	381	146	α = 32°, δ = + 61°............	5
107	1860	Nov.	7	7 0	9 0	....	46	....	N. E. of zenith..............	3
108	1860	"	12	6 30	17 0	14 30	423	68	....	
109	1860	"	12	10 0	16 0	14 30	381	90	In Leo...................	5
110	1860	"	13	6 15	17 0	13 30	500	83	....	.....
111	1860	"	13	15 15	16 15	....	21	....	In the "Sickle" in Leo	1
112	1860	"	14	15 40	16 25	....	15	....	" " "	1
113	1860	Dec.	12	8 20	12 0	11 30	180	60	....	.....
114	1861	April..	19	14 45	16 0	....	52	....	....	2
115	1861	July	3	9 5	10 5	....	36	....	....	4
116	1861	Aug.	10	8 0	15 15	13 30	397	75	β Persei	2
117	1861	"	10	10 0	13 0	12 30	289	119	Sword handle of Perseus......	4
118	1861	"	10	10 25	14 45	12 30	289	85	Between η, γ, and τ Persei ..	2
119	1861	"	10	10 30	13 0	....	95	....	In Perseus	1
120	1861	"	10	12 25	13 0	12 30	100	37	....	1
121	1861	"	11	8 15	10 15	....	52	....	....	3
122	1861	"	11	9 30	12 0	....	47	....	In Perseus	1
123	1861	Nov.	10	13 0	15 0	....	11	....	α = 150° 30′, δ = + 40° 40′ ..	1
124	1861	"	11	14 15	16 40	....	32	....	Near γ Leonis..............	1
125	1861	"	11	14 15	15 30	....	32	....	In Leo..................	2
126	1861	"	11	16 0	17 0	....	15	....	"	1
127	1861	"	12	13 0	17 0	16 30	130	72	"	4
128	1861	"	12	15 0	16 30	....	15	....	....	1
129	1861	"	12	11 15	12 15	....	11	....	....	4
130	1861	"	12	15 20	16 20	....	27	....	....	4
131	1861	"	13	10 0	13 0	....	15	....	Zenith................	1
132	1861	"	13	15 0	17 30	....	19	....	....	1
133	1861	"	13	15 15	17 38	....	23	....	....	1
134	1862	Aug.	9	14 0	15 40	....	90	....	In Perseus	2
135	1862	"	10	12 30	15 30	....	51	....	"	1
136	1862	Nov.	13	12 0	15 0	....	50	....	In Leo............	2
137	1862	"	13	15 15	17 5	....	31	....	"	1
138	1862	"	13	16 30	17 30	....	17	....	....	1
139	1863	Aug.	10	8 30	10 0	....	41	....	In Perseus........	1
140	1863	"	10	9 0	12 0	....	339	....	"	4
141	1863	"	10	9 0	13 0	....	257	....	....	2
142	1863	"	10	9 15	10 15	....	96	....	....	3
143	1863	"	10	10 0	13 50	....	130	....	....	1
144	1863	"	10	12 0	14 0	....	283	....	....	3
145	1863	"	10	13 0	14 0	....	87	....	....	1
146	1863	"	10	15 10	15 40	....	153	....	....	6
147	1863	"	11	9 0	11 0	....	67	....	....	1
148	1863	Nov.	11	10 0	15 0	....	105	....	....	.....
149	1863	"	11	11 22	14 52	....	185	....	....	.....
150	1863	"	12	10 0	14 0	....	129	....	....	.....
151	1863	"	12	10 20	15 51	....	109	....	....	.....
152	1863	"	13	8 0	14 0	....	107	....	....	8
153	1863	"	13	9 30	13 30	....	32	....	....	.....
154	1863	"	13	10 10	17 7	15 0	213	46	In Leo.................	6
155	1863	"	13	10 38	17 16	15 30	316	69	In the "Sickle" in Leo......	7

105. A. J. S. XXXI₂, 136.
106. " " "
107. A. J. S. XXXI₂, 137.
108. A. J. S. XXXI₂, 138.
109. A. J. S. XXXI₂, 139.
110. A. J. S. XXXI₂, 138.
111. A. J. S. XXXI₂, 137.
112. " " "
113. A. J. S. XXXI₂, 138.
114. A. J. S. XXXII₂, 294.
115. A. J. S. XXXII₂, 296.
116. A. J. S. XXXII₂, 294.
117. " " "

118. A. J. S. XXXII₂, 294.
119. A. J. S. XXXII₂, 447.
120. A. J. S. XXXII₂, 295.
121. A. J. S. XXXIII₂, 148.
122. A. J. S. XXXII₂, 447.
123. A. J. S. XXXIII₂, 148.
124. " " "
125. A. J. S. XXXIII₂, 146.
126. " " "
127. " " "
128. A. J. S. XXXIII₂, 147.
129. A. J. S. XXXIII₂, 148.
130. " " "

Meteor Showers–Cont'd.

Ref. number.	Authority.	Place of observation.	Remarks.
165	F. Bradley	Chicago, Ill.	
166	F. Bradley	Chicago, Ill.	
167	Prof. C. U. Shepard	Off Cape Hatteras.	
108	Francis Miller	Montgomery Co., Md	Several students assisted in the observations.
109	Prof. D. Kirkwood	Bloomington, Ind	Half the number conformable after 13h.
110	Francis Miller	Montgomery Co., Md	Several students assisted in the observations.
111	Prof. H. A. Newton	New Haven, Conn.	
112	Prof. H. A. Newton	New Haven, Conn.	
113	Francis Miller	Montgomery Co., Md	Several students assisted in the observations.
114	E. C. Herrick	New Haven, Conn.	No decided radiant.
115	E. C. Herrick	New Haven, Conn.	
116	F. W. Russell	Natick, Mass.	
117	E. C. Herrick	New Haven, Conn	A very large meteor at 11h. 30m.
118	B. V. Marsh	Burlington, N. J	A very large meteor at 11h. 23m.
119	Prof. A. C. Twining	New Haven, Conn.	
120	R. M. Gummere	Burlington, N. J.	
121	John Roberts	Madison, Ind.	
122	Prof. A. C. Twining	New Haven, Conn.	
123	F. W. Russell	Natick, Mass.	
124	F. W. Russell	Natick, Mass.	
125	Prof. A. C. Twining	New Haven, Conn.	
126	E. C. Herrick	New Haven, Conn.	
127	Prof. A. C. Twining	New Haven, Conn.	
128	S. J. Gummere	Burlington, N. J.	
129	Prof. D. Kirkwood	Bloomington, Ind.	
130	Prof. D. Kirkwood	Bloomington, Ind.	
131	F. W. Russell	Natick, Mass.	
132	S. J. Gummere	Burlington, N. J.	
133	B. V. Marsh	Germantown, Pa	Thirteen left trains.
134	F. W. Russell	Winchendon, Mass.	
135	B. V. Marsh	Germantown, Pa.	
136	Prof. S. J. Gummere	Haverford College, Pa.	
137	B. V. Marsh	Germantown, Pa.	
138	Prof. A. C. Twining	New Haven, Conn.	
139	W. G. Bryant	Winchendon, Mass.	
140	F. W. Russell	Natick, Mass.	
141	J. G. Pinkham	Manchester, Me.	
142	Prof. H. A. Newton	New Haven, Conn.	
143	B. V. Marsh	Philadelphia, Pa.	
144	F. W. Russell	Natick, Mass.	
145	F. Bradley	Chicago, Ill.	
146	Prof. H. A. Newton	New Haven, Conn.	
147	F. W. Russell	Natick, Mass.	
148	Prof. O. N. Stoddard	Oxford, Ohio.	
149	Prof. H. L. Smith	Kenyon College, Ohio.	
150	Prof. O. N. Stoddard	Oxford, Ohio.	
151	Prof. H. L. Smith	Kenyon College, Ohio.	
152	Prof. A. D. Bache	Coast Survey Office, Washington, D. C.	More than half the tracks plotted.
153	Prof. H. A. Newton	New Haven, Conn	Hazy.
154	Capt. J. M. Gilliss, U. S. N.	U. S. Naval Observatory.	Track of each meteor plotted.
155	Prof. S. J. Gummere	Haverford College, Pa	299 tracks plotted.

131. A. J. S. XXXIII₂, 148.
132. A. J. S. XXXIII₂, 147.
133.   "   "   "
134. A. J. S. XXXIV₂, 295.
135.   "   "   "
136. A. J. S. XXXV₂, 146.
137.   "   "   "
138.   "   "   "
139. A. J. S. XXXVI₂, 306.
140. A. J. S. XXXVI₂, 305.
141. A. J. S. XXXVI₂, 304.
142. A. J. S. XXXVI₂, 302.
143. A. J. S. XXXVI₂, 304.

144. A. J. S. XXXVI₂, 305.
145.   "   "   "
146. A. J. S. XXXVI₂, 302.
147. A. J. S. XXXVI₂, 305.
148. A. J. S. XXXVII₂, 144.
149. A. J. S. XXXVII₂, 146.
150. A. J. S. XXXVII₂, 144.
151. A. J. S. XXXVII₂, 146.
152. A. J. S. XXXVII₂, 145.
153. A. J. S. XXXVII₂, 143.
154. A. J. S. XXXVII₂, 141.
155. A. J. S. XXXVII₂, 142.

Ref. number.	Date.			Time of Shower.		Time of max. flight.	Whole number counted.	Max. hourly rate.	Radiant point.	No. of observers.
	Year.	Month.	Day.	Begin-ning.	End.					
				A. m.	A. m.	A. m.				
156	1863	Nov. ...	13	13 0	17 20	15 30	97	26	In the "Sickle" in Leo...	1
157	1863	"	13	15 0	16 30		79		" "	2
158	1863	"	13	15 45	17 45		49			1
159	1864	Aug.	9	9 0	18 0		17			1
159½	1864	"	9	10 0	11 0		33			1
160	1864	"	9	10 20	13 0		29			2
161	1864	"	9	10 30	13 0		332			4
162	1864	"	9	10 30	13 0		309			2
163	1864	"	9	10 40	12 0		13			2
163½	1864	"	9	11 30	14 30		44			2
164	1864	"	9	12 0	12 30		25			1
165	1864	"	9	13 0	15 20	14 30	694			3
166	1864	"	9	15 0	16 0		50			1
167	1864	Nov.	11	13 15	16 15		18		In Leo...	1
168	1864	"	12	13 45	16 0		46			1
169	1865	Aug.	9	14 25	15 50		19		α = 42°, δ = + 57°...	1
170	1865	"	12	10 45	12 45		16		α = 5?°, δ = + 56°...	1
171	1865	"	15	11 5	15 5	14 30	178	61		5
172	1865	Nov.	12	13 0	14 0		77		In the "Sickle" in Leo...	1
173	1865	"	12	15 0	18 0		63			1
174	1866	Aug.	9	9 0	9 20		14			1
175	1866	"	9	12 0	14 0	12 30	114	76		4
176	1866	"	10	9 0	10 15		50		In Cassiopeæ...	1
177	1866	"	10	9 10	15 0	14 30	364	97	In Perseus...	1
178	1866	"	10	9 15	11 30		160			1
179	1866	"	10	12 0	14 35		129			1
180	1866	"	10	12 0	14 15		164			1
181	1866	"	10	13 0	14 0		51			1
182	1866	"	10	14 0	16 0	14 30	73	43	In Perseus ...	1
183	1866	"	11	8 15	9 0		8			1
184	1866	"	11	14 0	16 0		63			1
185	1866	Nov.	8	9 45	13 20		93		"About Capella" ...	1
186	1866	"	9	10 0	15 0		47		In Leo...	1
187	1866	"	11	13 30	14 30		27			3
188	1866	"	12	7 0	18 0	14 30	205	45		1
189	1866	"	12	10 45	14 30	13 30	85	31		3
190	1866	"	12	11 0	16 40	16 10	402	22	Between ε and γ Leonis...	4
191	1866	"	12	11 10	13 40	11 30	236	108		12
192	1866	"	12	11 30	13 0		65			5
193	1866	"	12	12 0	17 0	15 30	603	155		15
194	1866	"	12	13 0	15 0		35			3
195	1866	"	12	13 10	13 40		8			1
196	1866	"	12	14 20	15 20		56			3
197	1866	"	12	14 40	17 30	17 15	458	190	α = 147° 30', δ = + 23° 15'...	12
198	1866	"	12	15 54	16 30		64			1
199	1866	"	12	16 6	16 36		8			7
200	1866	"	12				354			4
201	1866	"	13	7 0	18 0	15 30	453	71		
202	1866	"	13	8 30	16 30	15 0	440	138		4
203	1866	"	13	10 0	15 0		150			
204	1866	"	13	11 0	14 0	13 0	492	392		10
205	1866	"	13	11 0	16 10	15 30	901	212	In Leo...	12
206	1866	"	13	11 20	14 40	12 30	264	113		5

156. A. J. S. XXXVII₂, 142.
157. A. J. S. XXXVii₂, 143.
158. " " "
159. A. J. S. XXXVIII₂, 432
159½. " " "
160. " " "
161. " " "
162. " " "
163. " " "
163½. " " "
164. " " "
165. " " "
166. " " "
167. A. J. S. XXXIX₂, 229.

168. A. J. S. XXXIX₂, 229.
169. A. J. S. XL₂, 284.
170. " " "
171. " " "
172. A. J. S. XLI₂, 273.
173. " " "
174. A. J. S. XLII₂, 286.
175. A. J. S. XLII₂, 429.
176. A. J. S. XLII₂, 286.
177. A. J. S. XLII₂, 429.
178. " " "
179. " " "
180. " " "
181. A. J. S. XLII₂, 429.

Meteor Showers—Cont'd.

Ref. number.	Authority.	Place of observation.	Remarks.
156	B. V. Marsh	Germantown, Pa.	Hazy. *On Nov. 13, 55 tracks were plotted in Phila. by H. D.
157	C. E. Dutton	Norfolk, Va.	Vail, 56 tracks were plotted
158	J. H. Worrall	West Chester, Pa.	Hazy. in West Chester by B. Hoop-
159	H. P. Tuttle	Off Charleston, S. C.	er, 27 tracks were plotted in
160	G. Scarborough	Riverside, Kan.	Easton by E. Menline, 23
160	H. S. Osborn	Belvidere, N. J.	tracks were plotted in St. Louis by Prof. W. Chauvenet.
161	Francis Bradley	Chicago, Ill.	Cloudy in the Middle and N. E. States.
162	W. H. R. Lykins	Lawrence, Kan.	
163	B. V. Marsh	Philadelphia, Pa.	
163	Prof. H. A. Newton	New Haven, Conn.	
164	G. Scarborough	Riverside, Kan.	
165	Francis Bradley	Chicago, Ill.	
166	G. Scarborough	Riverside, Kan.	
167	R. H. Stretch	Virginia City, Nevada.	
168	R. H. Stretch	Virginia City, Nevada.	
169	Prof. A. C. Twining	Hinsdale, Mass.	
170	Prof. A. C. Twining	Hinsdale, Mass.	
171	Prof. H. A. Newton	New Haven, Conn.	
172	O. B. Wheeler	Detroit, Mich.	
173	O. B. Wheeler	Detroit, Mich.	
174	D. Trowbridge	Hector, N. Y.	
175	Prof. H. A. Newton	Sherborne, N. Y.	
176	D. Trowbridge	Hector, N. Y.	Some left trains lasting 6 or 7 seconds.
177	F. W. Russell	Natick, Mass.	
178	Isaac Pierson	Vineyard Sound	Off Martha's Vineyard, Mass.
179	R. V. Marsh	Germantown, Pa.	
180	R. M. Gammere	Germantown, Pa.	
181	Prof. H. A. Newton	Sherborne, N. Y.	
182	J. H. Worrall	West Chester, Pa.	
183	D. Trowbridge	Hector, N. Y.	
184	J. H. Worrall	West Chester, Pa.	
185	James Ferguson	U. S. Naval Observatory.	
186	James Ferguson	U. S. Naval Observatory.	
187	F. Bradley	Chicago, Ill.	
188	Prof. Hopkins	Williamstown, Mass.	Williams College.
189	C. G. Rockwood	Newark, N. J.	No observations from **12h. 45m. to** 13h. 0m.
190	James Ferguson	U. S. Naval Observatory.	
191	Prof. H. A. Newton	New Haven, Conn.	
192	F. Bradley	Chicago, Ill.	F. W. Russell, of Cambridge, Mass.,
193	C. S. Lyman	New Haven, Conn.	observed several evenings in first
194	Prof. Hinrichs	Iowa City, Iowa.	half of November, and counted, in
195	B. V. Marsh	Germantown, Pa.	all, 875 meteors.
196	O. B. Wheeler	Detroit, Mich.	
197	Prof. H. A. Newton	New Haven, Conn.	
198	Prof. D. Kirkwood	Canonsburg, Pa.	
199	B. V. Marsh	Germantown, Pa.	
200	Maria Mitchell	Vassar College, N. Y.	Observed seven hours.
201	Prof. Hopkins	Williamstown, Mass.	Williams College.
202	Prof. Hinrichs	Iowa City, Iowa.	
203	John T. Wheeler	Concord, N. H.	
204	Prof. C. S. Lyman	New Haven, Conn.	
205	Prof. H. A. Newton	New Haven, Conn.	
206	C. G. Rockwood	Newark, N. J.	

182. A. J. S. XLII₂, 429.
183. A. J. S. XLII₂, 286.
184. A. J. S. XLII₂, 429.
185. A. J. S. XLII₂, 78; Nov. Meteors, 1866, U. S. N. Obs'y, 8vo.
186. "   "   "   "   "
187. A. J. S. XLIII₂, 78.
188. "   "   "   "
189. "   "   "
190. A. J. S. XLIII₂, 78; Nov. Meteors, 1866, U. S. N. Obs'y, 8vo.
191. A. J. S. XLIII₂, 78.
192. "   "   "
193. "   "   "

194. A. J. S. XLIII₂, 78.
195. "   "   "   "
196. "   "   "   "
197. "   "   "   "
198. "   "   "   "
199. "   "   "   "
200. "   "   "   "
201. "   "   "   "
202. "   "   "   "
203. Smithsonian Archives (not yet printed).
204. A. J. S. XLIII₂, 78.
205. "   "   "   "
206. "   "   "   "

*A. J. S. XXXVII₂, 141.

CATALOGUE IV.—

Ref. number	Year.	Month.	Day.	Time of Shower. Beginning.	End.	Time of max. flight.	Whole number counted.	Max. hourly rate.	Radiant point.	No. of observ's.
				A. m.	A. m.	A. m.				
207	1866	Nov. ...	13	12 0	12 36	....	28	....	....	1
208	1866	" ...	13	12 0	13 0	....	35	....	a = 147° 30', δ = + 24° 30'.....	1
209	1866	" ...	13	12 0	17 15	14 30	101	51	....	
210	1866	" ...	13	14 0	16 0	....	180	....	....	3
211	1866	" ...	13	14 0	16 30	14 30	173	91	e Leonis............................	4
212	1866	" ...	13	15 8	16 8	....	43	....	Bend of the Sickle............	1
213	1866	" ...	13			....	419	....	....	6
214	1866	" ...	16	13 0	15 0	....	24	....	Between μ and e Leonis........	1
215	1866	" ...	17	14 0	17 30	....	75	....	Near a Leonis....................	1
216	1867	May....	20			....	20	....	A little southeast of Lyra....	1
217	1867	Aug....	8	12 15	12 55	....	17	....	In Perseus	1
218	1867	" ...	9	13 3	14 30	....	35	....	"	1
219	1867	" ...	9	13 3	14 30	....	27	....	"	1
220	1867	" ...	9	14 15	15 25	....	39	....	"	1
221	1867	" ...	10	14 0	15 40	....	44	....	"	1
222	1867	" ...	10	15 0	15 30	....	17	....	"	1
223	1867	" ...	11	14 0	15 30	....	45	....	"	1
224	1867	Nov. ...	13	9 0	17 0	15 30	535	276	....	5
225	1867	" ...	13	9 28	13 0	....	50	....	....	1
226	1867	" ...	13	10 0	16 0	....	4,937	....	....	10
227	1867	" ...	13	11 0	16 0	....	1,000	....	....	
228	1867	" ...	13	11 45	18 25	15 30	2,325	1,488	....	
229	1867	" ...	13	12 0	16 40	15 45	2,056	....	....	4
230	1867	" ...	13	12 0	17 45	15 50	3,730	2,184	....	
231	1867	" ...	13	12 30	17 0	15 52	1,713	1,076	a = 147° 45', δ = + 23° 0'....	2
232	1867	" ...	13	13 0	18 0	16 30	3,044	2,110	γ Leonis..............................	
233	1867	" ...	13	13 0	13 30	....	500	....	....	
234	1867	" ...	13	13 0	16 28	16 22	1,474	2,220	In Leo................................	15
235	1867	" ...	13	13 0	18 0	16 10	2,267	1,472	....	4
236	1867	" ...	13	14 0	18 13	15 30	1,626	1,082	....	4
237	1867	" ...	13	15 5	17 45	....	1,800	....	In Leo................................	1
238	1867	" ...	13	15 45	16 45	....	1,088	....	a = 150° 45', δ = + 21° 55'	1
239	1867	" ...	13	15 55	16 35	....	339	....	a = 148° 0', δ = + 23° 0'	2
240	1867	" ...	13	16 0	17 0	....	500	....	....	2
241	1867	" ...	13	16 5	17 54	....	1,361	....	In Leo................................	4
242	1867	" ...	13	16 30	17 30	....	460	....	....	1
243	1867	" ...	13	17 0	17 22	....	100	....	In Leo................................	1
244	1867	" ...	13	17 0	17 30	....	61	....	....	1
245	1867	" ...	13			....	500	....	....	1
246	1868	Aug....	9	13 15	14 0	....	56	....	In Perseus	2
247	1868	Nov. ...	13	10 0	15 30	....	1,350	....	....	3
248	1868	" ...	13	10 45	18 0	16 30	2,886	572	....	2
249	1868	" ...	13	11 0	18 11	....	2,500	....	....	
250	1868	" ...	13	11 30	17 30	....	1,462	....	a = 152°, δ = + 18°............	1
251	1868	" ...	13	11 35	13 20	....	700	....	....	3
252	1868	" ...	13	12 0	17 54	17 27	5,573	1,402	In Leo................................	10
253	1868	" ...	13	11 34	17 40	17 0	5,000	1,400	....	
254	1868	" ...	13	12 28	16 20	....	800	....	....	2
255	1868	" ...	13	12 51	17 15	17 10	5,670	1,543	....	
256	1868	" ...	13	13 0		....	896	....	....	1
257	1868	" ...	13	13 0	17 28	....	850	....	....	1
258	1868	" ...	13	13 0	18 0	16 30	1,341	828	....	4
259	1868	" ...	13	13 4	17 37	....	1,926	....	....	3

207. A. J. S. XLIII₂, 78.
208. "   "   "
209. A. J. S. XLIII₂, 413.
210. A. J. S. XLIII₂, 78.
211. A. J. S. XLIII₂, 78; Nov. Meteors, 1866, U. S. N. Obs'y, 8vo.
212. A. J. S. XLIII₂, 78.
213.
214. A. J. S. XLIII₂, 78; Nov. Meteors, 1866, U. S. N. Obs'y, 8vo.
215. "   "   "
216. Trans. St. Louis Acad. Sci., II, 577.
217. A. J. S. XLIV₂, 426.
218. "   "   "
219. "   "   "

220. A. J. S. XLIV₂, 426.
221. "   "   "
222. "   "   "
223. "   "   "
224. A. J. S. XLV₂, 78.
225. "   "   "
226. A. J. S. XLV₂, 225.
227. A. J. S. XLV₂, 78.
228. Smithsonian Archives (not yet printed).
229. A. J. S. XLV₂, 225.
230. A. J. S. XLV₂, 78.
231.
232. Proc. Am. Phil. Soc. X, 356.
233. A. J. S. XLV₂, 78.
234. "   "   "

Meteor Showers—Cont'd.

Ref. number.	Authority.	Place of observation.	Remarks.
207	B. V. Marsh	Germantown, Pa.	
208	Prof. A. C. Twining......	New Haven, Conn.	
209	The Denver News......	Denver, Col.	
210	W. A. Anthony.............	Franklin, N. Y.	
211	James Ferguson..........	U. S. Naval Observatory.	
212	Prof. A. C. Twining....	New Haven, Conn.	
213	Maria Mitchell............	Vassar College, N. Y.	
214	James Ferguson..........	U. S. Naval Observatory......	120 tracks plotted on charts from Nov.
215	James Ferguson..........	U. S. Naval Observatory.	9 to Nov. 17, inclusive.
216	R. Hayes	St. Louis, Mo...................	Observations made in the evening.
217	F. W. Russell..............	Winchendon, Mass.............	"Moon two days past the full."
218	B. V. Marsh...............	Philadelphia, Pa.	
219	C. H. Darlington.........	Philadelphia, Pa.	
220	Lewis Swift................	Marathon, N. Y.	
221	F. W. Russell..............	Winchendon, Mass.	
222	Lewis Swift................	Marathon, N. Y.	
223	F. W. Russell..............	Winchendon, Mass.	
224	Prof. T. C. Wylie.........	Bloomington, Ind.	
225	Prof. H. A. Newton.....	New Haven, Conn.	Cloudy from 12h. 0m. to 15h. 15m.
226	Prof. N. R. Leonard....	Iowa City, Iowa.................	
227	S. J. Gummere	Haverford College, Pa.	
228	Prof. F. H. Snow.........	Lawrence, Kan.	
229	Prof. C. A. Young.......	Hanover, N. H...................	Dartmouth College.
230	Prof. T. H. Safford......	Chicago, Ill.	Number of observers varied from 8
231	Francis Bradley..........	Evanstown, Ill.	to 30.
232	James McClure	Philadelphia, Pa	Several observers.
233	San Francisco Times...	San Francisco, Cal.	
234	Prof. H. A. Newton.....	New Haven, Conn...............	One-third of the sky covered with
235	G. F. Kingston	Toronto, Canada.	clouds.
236	Robert B. Taber..........	New Bedford, Mass.	
237	Profs. Newcomb and Eastman.	U. S. N. Observatory	147 tracks mapped.
238	Prof. J. C. Watson......	Ann Arbor, Mich.	
239	Prof. W. Harkness	Richmond, Va.	
240	Mrs. J. H. Trumbull...	Hartford, Conn.	
241	Prof. G. W. Hough	Dudley Observatory	*In Chihuahua, Mexico, the meteors
242	J. N. Flint...................	Canaseraga, N. Y.	fell so fast they could not be counted;
243	Prof. A. C. Twining.....	New Haven, Conn.	often 20 or 30 visible at once.
244	J. D. Parker...............	Topeka, Kan.	
245	Prof. E. Loomis	New Haven, Conn...............	Observed one hour.
246	C. G. Rockwood	Durham, Conn.	
247	Robert B. Taber..........	New Bedford, Mass	Actual observing time 4h. 35m.
248	G. T. Kingston	Toronto, Canada.	
249	T. A. Wylie.................	Bloomington, Ind.	
250	W. S. Gilman, Jr........	Palisades, N. Y...................	†Several bright meteors were seen
251	R. J. Gilman...............	Williamstown, Mass.	during the Nov. shower in 1868.
252	Prof. H. A. Newton	New Haven, Conn.	
253	S. J. Gummere	Haverford College, Pa.	
254	Prof. J. R. Eastman ...	U. S. N. Observatory.	
255	C. G. Rockwood	Brunswick, Me.	
256	Lewis Swift................	Marathon, N. Y...................	The time when observations ended
257	P. E. Chase.................	Haverford College, Pa.	not given.
258	W. H. Pratt.................	Davenport, Iowa	43 were counted before 13h. 0m.
259	C. G. Berner...............	Veray, Ind.	

235. A. J. S. XLV₂, 78.
236. "   "   "
237. A. J. S. XLV₂, 225; Nov. Meteors, 1867, U. S. N. Obs'y, 8vo.
238. A. J. S. XLV₂, 78.
239. A. J. S. XLV₂, 225; Nov. Meteors, 1867, U. S. N. Obs'y, 8vo.
240. A. J. S. XLV₂, 78.
241. "   "   "
242. "   "   "
243. "   "   "
244. "   "   "
245. "   "   "
246. A. J. S. XLVII₂, 297.
247. A. J. S. XLVII₂, 118.
248. "   "   "

249. A. J. S. XLVII₂, 118.
250. "   "   "
251. "   "   "
252. "   "   "
253. "   "   "
254. Nov. Meteors, 1868, U. S. N. Obs'y, 8vo.
255. A. J. S. XLVII₂, 118.
256. "   "   "
257. A. J. S. XLVII₂, 118; Proc. Am. Phil. Soc. X, 539.
258. A. J. S. XLVII₂, 118; Dav. Acad. Nat. Sci. I, 14.
259. A. J. S. XLVII₂, 118.
 *A. J. S. XLV₂, 78.
 †A. J. S. XLVII₂, 399.

CATALOGUE IV.-

Ref. number.	Date.			Time of Shower.		Time of max. flight.	Whole number counted.	Max. hourly rate.	Radiant point.	No. of observ's.
	Year	Month.	Day.	Beginning.	End.					
				h. m.	h. m.	h. m.				
260	1868	Nov	13			14 30	3,766	900		5
261	1868	"	13	14 20	17 58	16 55	4,278	1,650		4
262	1868	"	13	15 0	17 0		1,780			3
263	1868	"	13	15 7	17 38		455			1
264	1868	"	13	15 30	18 0		711			1
265	1868	"	13	16 0	17 15		250			1
266	1868	"	13	16 30	17 30		833			2
267	1869	"	13	7 0	19 0		830			4
268	1869	"	13	13 18	15 43		556			2
269	1870	"	12	12 30	15 30		31		In Leo	3
270	1870	"	13	11 5	15 45		153			6
271	1870	"	13	13 20	16 10		82		"	3
272	1870	"	13	13 30	15 0		13			1
273	1870	"	13	13 30	15 0		11			1
274	1871	Aug	10	11 40	13 18		283			6
275	1871	Nov	13				98			
276	1872	Aug	9	10 30	15 30		358			4
277	1872	"	9	14 45	15 55		62			1
278	1872	Nov	24	7 30	12 30		217		2° or 3° N. of γ Andromedæ	2
279	1872	"	25	10 25	14 0		80			5
280	1872	"	27	6 5	10 0	6 30	720	300	μ Andromedæ	1
281	1872	"	27	6 10	11 45		143		$\alpha = 15^\circ, \delta = +30^\circ$	1
282	1872	"	27	6 38	8 48	6 55	1,730	1,300	$\alpha = 25^\circ, \delta = +43^\circ$	4
283	1875	Aug	5	13 30	15 30		112		In Perseus	1
284	1876	Oct	24	11 30	15 30		58		Bet. Orion and the Pleiades	1
285	1876	Nov	13	12 0	13 0		12			1
286	1877	"	13	13 55	15 45		54		In Leo	2
287	1885	"	27	6 0	6 24		44			1
288	1885	"	27	9 0			12		γ Andromedæ	3
289	1885	"	27	10 15	10 30		13			1
290	1885	"	27	7 15	7 45		100		2° N. W. of γ Andromedæ	1
291	1885	"	27	6 30	7 50		213			3
292	1885	"	27	7 0	9 0		328			4

260. A. J. S. XLVII$_2$, 118.
261. Nov. Meteors, 1868, U. S. N. Observatory, 8 vo.
262. A. J. S. XLVII$_2$, 118.
263. " " "
264. " " "
265. " " "
266. " " "
267. A. J. S. XLIX$_2$, 244.

268. A. J. S. XLIX$_2$, 244.
269. A. J. S. I$_3$, 30.
270. " " "
271. " " "
272. " " "
273. " " "
274. A. J. S. II$_3$, 227.
275. A. J. S. II$_3$, 470.
276. A. J. S. IV$_3$, 244.

Meteor Showers—Cont'd.

Ref. number.	Authority.	Place of observation.	Remarks.
260	Maria Mitchell	Vassar College, N. Y.	
261	Prof. J. R. Eastman ...	U. S. N. Observatory............	90 tracks plotted.
262	B. J. Gilman................	Williamstown, Mass.	
263	C. W. Tuttle................	Boston, Mass.....................	No observations between 16h. 4m. and
264	Robert B. Taber..........	New Bedford, Mass.	16h. 33m.
265	J. E. Hendricks..........	Des Moines, Iowa.	
266	B. F. Mudge................	Manhattan, Kan.	
267	....	Fredericton, N. B.............	Reported by Prof. H. A. Newton.
268	George Davidson........	Santa Barbara, Cal.	
269	Prof. H. A. Newton......	New Haven, Conn.	
270	Prof. H. A. Newton	New Haven, Conn.	
271	C. G. Rockwood	Brunswick, Me..................	6 tracks plotted.
272	B. V. Marsh................	Burlington, N. J.	
273	J. G. Gummere	Burlington, N. J.	
274	Prof. R. A. Newton.....	Sherburne, N. Y.	
275	Prof. H. A. Newton......	New Haven, Conn.............	Only a few belonged to the November
276	Prof. C. A. Lyman......	New Haven, Conn.	system.
277	R. W. McFarland	Oxford, Ohio.	
278	Prof. H. A. Newton......	New Haven, Conn.	
279	Prof. H. A. Newton......	New Haven, Conn.	
280	Prof. J. R. Eastman ...	U. S. N. Observatory.	
281	B. V. Marsh................	Philadelphia, Pa.	
282	Prof. H. A. Newton......	New Haven, Conn.	
283	Prof. J. R. Eastman ...	U. S. N. Observatory............	Most of the meteors had short trains.
284	Prof. J. R. Eastman	U. S. N. Observatory............	A few left long trains.
285	Prof. H. A. Newton......	New Haven, Conn	Night mostly cloudy; 3 conformable
			to radiant.
286	Prof. D. Kirkwood	Bloomington, Ind	Night mostly cloudy.
287	Robert Brown............	New Haven, Conn.	
288	H. A. Newton.............	New Haven, Conn	Hourly number estimated for 3 ob-
289	H. A. Newton.............	New Haven, Conn.	servers, 100.
290	C. A. Young	Princeton, N. J.	
291	A. Hall	Georgetown, D. C.	
292	D. Horigan................	Georgetown, D. C.	

277. A. J. S. IV₃, 244.
278. A. J. S. V₃, 53.
279. "   "   "
280. "   "   "
281. "   "   "
282. "   "   "
283. Astron'l and Meteor'l Observ'ns, 1875.
284. Astron'l and Meteor'l Observ'ns, 1876.

285. A. J. S. XII₂, 473.
286. A. J. S. XV₃, 76.
287. A. J. S. XXXI₃, 78.
288. "   "   "   "
289. "   "   "   "
290. "   "   "   "
291. "   "   "   "
292. "   "   "   "

CATALOGUE V.—Sporadic Meteors.

Ref. number.	Year.	Month.	Day.	Hour and minute. (h. m. s.)			Place of observation.	Apparent size.	Color.	Duration.	Position, altitude and azimuth.
1	1779	Oct.	31	6	10	0	Williamsburg, Va.			s.	Fell and disappeared below the horizon 4° N. of W.
2	1813	Mar.	21	19	0	0	New Haven, Conn.	Nearly as large as the full moon.	Like the moon	8 or 16	Alt., 35°; az., N. 29° E. Alt., 30°; az., N. 45° E.
3	1819	July	24	Evening			Youngstown, Ohio.				
4	1819	Nov.	21	10	0	0	Easton, Penn.				
5	1822	Mar.	9	7	30	0	New York, N. Y.	One-third size of the moon.		5	Alt., 48°; az., N. 45° E.
6	1826	Mar.	31				New Haven, Conn.	1' in diameter	White tinged with blue.		Alt., 43°; az., S. 13° E. Alt., 25°; az., S. 55° W.
7	1826	Apr.	1	9	0	0	New Haven, Conn.	At appearance 7' and at disappearance 28' in diameter.		9.5	Alt., 66°; az., N. 30° E.
8	1826	Apr.	11	11	20	0	Burlington, Vt.				Alt., 9° 45'; az., N. 41° 54' E. Alt., 3° 6'; az., N. 26° 57' E.
9	1828	Feb.	11	11	12	0	On the East river, at New York.		A light shade of "grass green."	2	
10	1836	Aug.	30	4	0	0	Eight miles S. of Jacksonville, Ill.	Diameter 18'.	Like a globe of fire.		Alt., 60°; az., N. 18° E. Alt., 40°; az., N. 15° E.
11	1837	July	5	7	8	0	Rochester, N. H.	As large as the sun.	Like "white-heated iron."	1m.	Appeared in the S. W. at an altitude of 30°.
12	1837	Dec.	14	7	29	32	New Haven, Conn.	From ¼ to ¾ the size of the moon.		1.0 or 1.5	Disappeared at altitude of 9°; azimuth, S. 89° W.
13	1838	May	18	8	0	0	Throughout the northern United States and part of upper Canada.	One-fourth the diameter of the moon.		7	Alt. at beginning of path, 28.5 miles; at middle of path, 34.8 miles; at end of path, 32.1 miles.
14	1840	May	12	15	0	0	Four miles south of Stonington, Conn.	Larger than the full moon.		3m. 30s.	Appeared in the N. E., and disappeared in the N. N. E., near the horizon.

1. Trans. Am. Phil. Soc. II (O. S.), 173, 175.
2. A. J. S. XIII., 33.
3. A. J. S. VI., 315.
4. A. J. S. VI., 316.
5. A. J. S. VI., 318.
6. A. J. S. XI., 184.
7.

8. A. J. S. XI., 120.
9. A. J. S. XIV., 199.
10. A. J. S. XXXVIII., 402.
11. A. J. S. XXXVIII., 200.
12. A. J. S. XXXVII., 120.
13. A. J. S. XXXIV., 223.
14. A. S. S. XXXIX., 382.

CATALOGUE V.—Sporadic Meteors—Continued.

Ref. number.	Train, appearance and duration.	Length of path. °	Direction of motion.	Remarks.	Observer.
1	Train visible 15m			Seen in Philadelphia by David Rittenhouse	John Page.
2	Train 12° long		Easterly nearly horizontal	The light was powerful; cast deep shadows. In 8 or 10 minutes after its disappearance a heavy report, accompanied by a decided jar, was heard.	S. E. Dwight.
3			Moving towards the north	Disappeared with an explosion, the report of which was heard in about three minutes. At Gustavus, Ohio, the same object was seen and the report heard in about one minute.	Dr. H. Manning.
4			Towards the S. or S.W	Disappeared at an altitude of 35° with a loud explosion, the report of which was heard in "two, three, or four minutes."	Samuel Turney.
5				This meteor was seen in Canada, throughout New England and New York, and in several places in Pennsylvania. At Troy, N. Y., the report of the explosion was heard in 7m. 30s.	
6	Train several degrees long			Report of explosion heard in 4m. 25s.	
7	Train visible one minute		Disappeared at the zenith	Report heard in two minutes; also seen at Arlington, Vt., where it seemed to move from north to south, and vanished at an altitude of 30°. Report heard in 1m. 30s.	
8				No explosion.	
9	Train of same color as the meteor	20	N. E. from a point 55° "below the zenith."	The meteor seemed to burst into fragments, but no report was heard.	Dr. H. S. Waterhouse. Prof. B. D. Silliman.
10	Left a "cloud of smoky appearance."	20	Vertically towards the earth	Report of explosion at disappearance distinctly heard.	R. Gaylord.
11	Left a greyish train which, after waving and expanding for several minutes, gradually vanished.		Course towards the N. N.W. and inclined to the horizon.		
12	Brilliant train, visible a long time.		Downward at an angle of from 30° to 50° with the vertical.	Seemed to explode, but no report was heard. Position at appearance not well observed.	E. C. Herrick.
13	A brilliant train broken into several sections.		Nearly north and horizontal	Data collected and computations made by Prof. E. Loomis.	Profs. Hubbard, Catlin, and Barrows, and Messrs. R. W. Haskins, E. Brown, and J. and L. Chandler.
14	Brilliant train lasting several seconds after the extinction of the meteor.		Downward at an angle of 30° with the horizon.	Exploded with a loud report, which was heard in ten or fifteen seconds.	J. J. Comstock.

CATALOGUE V.—Sporadic Meteors—Continued.

Ref. number	Year	Month	Day	Hour and minute	Place of observation	Apparent size	Color	Duration	Position, altitude and azimuth.
15	1841	Nov.	10	8h. 0m.	Boston, Mass.		Preceding portion white; the following blue.	* .	Passed several degrees west of Famalhaut and disappeared at an altitude of 10°.
16	1841	Nov.	10	8 0	New Haven, Conn.	Diameter 3' or 4'.		2 or 3	Appeared 30' below the Pleiades and disappeared midway between β and ρ Ceti.
17	1845	Aug.	11	6 5	Cambridge, Mass.	Estimated diameter less than 5'.	White.	1	Alt., 25° 30'.
18	1845	Aug.	25	8 0	Cambridge, Mass.	Half the diameter of the full moon.	Red and blue.		Alt., 45°; az., S. 10° W.
19	1845	Aug.	31	14 20	Fayetteville, N. C.	"As large as the sun."			
20	1846	July	13	9 30	York and Lancaster, Pa.	From one-third to two-thirds the size of moon.		20	Alt., 10°; az., S. 60° W. Alt., 8°; az., N. 88° W.
21	1846	July	20	9 55	Cambridge, Mass.	"Equal to Venus".	Red and yellowish white.		
22	1846	Aug.	11	10 10	Cambridge, Mass.	"Brighter than Venus".	Blue, red, and white.	2	Alt., 35°; az., N. 85° W. Alt., 39°; az., N. 63° W.
23	1848	Jan.	20	5 0	Rome, N. Y.				Alt., 36°.
24	1848	Mar.	5	14 30	Nantucket, Mass.	39' in diameter.			Alt., 31°; az., N. 12° E. Alt., 16°; az., N. 12° W.
25	1848	April	15		Whitesville, Miss.				
26	1850	June	16	6 25	New Haven, Conn.	"As large as Venus".	Brilliant white.		Exploded and disappeared in a 2h. 30m. and δ + 27°.
27	1850	Sept.	30	8 54	Boston, Mass.				
28	1850	Sept.	30	8 53	Springfield, Mass.		Bluish.	From 60 to 90m.	Alt., 64°; az., S. of W.
29	1850	Oct.	3	8 39	Hartford, Conn.		Orange.		
30	1856	July	8	6 6	Thornhill, Hancock Co., Ala.	"6 or 8 inches in diam'r".			First seen at a point midway between the zenith and horizon; az., S. 30° or 35° W.

15, A. J. S. XLIII., 399.
16, "
17, Proc. Am. Acad. Arts and Sci. I., 71.
18, "
19, A. J. S. XLIX., 408.
20, A. J. S. XLI., 347.

21, Proc. Am. Acad. Arts and Sci. I, 22.
22, A. J. S. V., 457.
23, Proc. Am. Acad. Arts and Sci. I, 329.
24, A. J. S. VI., 149.

25, A. J. S. XI., 131.
26, " "
27, " "
28, A. J. S. XXIII., 448; XXIII., 138, 237.

CATALOGUE V.—Sporadic Meteors—Continued.

Ref. number.	Train, appearance and duration.	Length of path.	Direction of motion.	Remarks.	Observer.
		°			
15	Left a train 2° or 3° long near and a little above β Aquarii, visible more than 90s.		Towards the S. W.		W. C. Bond and Son.
16	Train 8° or 10° long and visible 3m.	30		Same meteor as the preceding.	W. C. Bond.
17		7	Toward the horizon at an angle of 50°.		W. C. Bond.
18		10 or 20	From N. E. to S. E.		W. C. Bond.
19				Exploded with a very loud report, which was heard in five or six minutes.	D. M. Ettinger.
20	Train 1° long, seen only at York.		From γ Cygni to near a Cassiopeæ.	Seen also at New Haven and many other places.	W. C. Bond.
21			From a Cassiopeæ to 80 Pegasi.		W. C. Bond.
22	A cometary tail of dense white light.				W. C. Bond.
23	Brilliant train, a portion visible 12m. or 18m.			Light of train could be seen through light cirrus clouds.	Edward Huntington.
24				Loud report heard in 92s. after the meteor disappeared over the ocean.	William Mitchell.
25	Left a bright train.		From a little W. of N. towards the E.	Disappeared with a loud explosion. Light strong enough to cast deep shadows, even in the moonlight.	
26	Train 1° long.			At the time of appearance the sun was about 15° above the horizon and unobserved.	George Rice and J. P Humaston.
27	Left a brilliant train extending from the head of Medusa, to 3° below a Arietis, visible more than one hour.			The train was examined with the Cambridge equatorial, and it presented the appearance of minute bright clouds resembling cirro-cumuli.	W. C. Bond.
28	The train was yellowish, and remained stationary for about 4m. It then slowly changed its form and its direction of motion, and at 9h. 18m. it was barely visible as a faint white cloud in the lower portion of Aries.		Horizontally from E. to W.	Same meteor as the preceding.	J. C. Hoadley.
29	Train 15 or 16 feet long.		From W. to S. E.	Disappeared below S. E. horizon. No explosion. Train swept over the moon's disk. Seen also at Columbus, Grenada, and Holly Springs, Miss., and at Marion, Ala.	Gaylord Wells, by Prof. Brocklesby. T. M. Peters and B. R. Delgraffenried.
30	White train, visible 15m.	35	Nearly vertical.		

CATALOGUE V.—Sporadic Meteors—Continued.

Ref. number.	Year	Month	Day	Hour and minute.	Place of observation.	Apparent size.	Color.	Duration.	Position, altitude and azimuth.
31	1857	April	11	A.M. 8 50	Near Lake Winnibigoshish, Minn.	"Greater than the full moon."		s. 4	Appeared in Hydra and moved W. to within a short distance of the horizon.
32	1859	Aug.	11	7 20	From Newburyport, Mass. to Petersburg, Va.	"As large as the sun"	Red.		
33	1859	Nov.	14	21 30		Nearly as large as the sun.	Like a ball of fire	2	Exploded over the vicinity of Cape May.
34	1860	July	20		From Lake Michigan to Nantucket, Mass., and from Maine to Virginia.	Nearly as large as the moon.			Computed altitudes: Over Lake Michigan, 120 miles. " Lake Huron, 85 miles. " Buffalo, N. Y., 62 miles. " Elmira, N. Y., 51 miles. " Long Island Sound, 42 miles.
35	1860	Aug.	2	Bet'n 10 and 11h.	Knoxville, Tenn.				
36	1860	Aug.	6	Bet. 7.50 and 8h.	New York and New Haven, Conn.				
37	1861	Aug.	10	11 30	New Haven, Conn.			1 or 2	Appeared 1° S.W. of ε Cygni and passed near the Dolphin.
38	1861	Oct.	4	7 29	New Haven, Conn.				
39	1861	Dec.	8	13 0	Brunswick, Me.				
40	1861	Dec.	17	4 45	Buffalo, N. Y.				
41	1861	Dec.	25	"Sunset"	Southern Connecticut.				In the S. E.
42	1861	Dec.	30	19 0	Hartford, Conn.				Over the Atlantic Ocean.
43	1862	Jan.	2	19 17	New York, N. Y.				
44	1862	Jan.	5	"Sunset"	Setauket, N. Y.	One-third diam'r of moon.	"Sparkling white."		
45	1862	Aug.	13	7 47	U. S. Naval Observatory.				
46	1862	Sept.	3	12 12	U. S. Naval Observatory.			3	Near Pleiades.
47	1862	Oct.	3	12 15	U. S. Naval Observatory.			4	In the N.
48	1862	Nov.	3	7 0	U. S. Naval Observatory.				
49	1862	Dec.	11	7 0	U. S. Naval Observatory.				

31. A. J. S. XXIV, 168.
32. A. J. S. XXVIII, 309.
33. { A. J. S. XXIX, 137, 298, 447. Jour. Frank. Inst. XXXIX, 256, 258. { A. J. S. XXX, 186.
34. A. J. S. XXX, 293; Smithsonian Contributions, XVI.
35. A. J. S. XXX, 295.
36. A. J. S. XXX, 296; XXXII, 346.
37. A. J. S. XXXII, 48.

38. A. J. S. XXXII, 443.
39. A. J. S. XXXIII, 224.
40. " " "
41. " " "

CATALOGUE V.-Sporadic Meteors—Continued.

Ref. number.	Train, appearance and duration.	Length of path.	Direction of motion.	Remarks.	Observer.
31	White train lasting 5m. or 10m.	°	Nearly horizontal	Seen in New York, Connecticut, and Vermont. At its disappearance the explosion jarred the buildings in Troy, N. Y. The sun was shining brightly.	B. F. Odell.
32					
33	No luminous train was seen, but a trail of smoke was left in its path.	15 or 20	Nearly vertical	Position computed by Prof. E. Loomis.	
34				In the vicinity of Elmira, N. Y., the meteor separated by explosion into two parts, which continued their course until over Nantucket, when by another explosion they were separated into small fragments, which continued their course as before.	Positions computed by Prof. C. S. Lyman.
35			From E. to W.	Seen throughout the southern portion of the United States.	W. C. Kane.
36			From S. to N.		Prof. A. C. Twining.
37	Brilliant train, visible more than 15s.	29	Slightly inclined towards the earth.	Also seen at Burlington, N. J., by B. V. Marsh.	
38	Brilliant train reaching from Cygnus to Cassiopeia. It slowly assumed a serpentine shape and remained visible 2m.		From west to east nearly through the zenith.		
39			From W. to E.		
40	Left a short train.			Broke into three fragments.	
41			From an altitude of 10° obliquely to horizon.	Exploded with a report like a pistol.	
42			From a Cygni to Delphinus.		
43			From E. to W.		
44			From the N. E.		
45			From N. W. to S. E.		
46			From E. to W.		
47	Train visible 10s.		From N. E. to N. W.		
48			From zenith to S. W.		
49					

42. A. J. S. XXXIII., 291.
43. " "
44. " "
45. Wash'n Ast. & Met. Obs'ns, 1862.
46. " "
47. " "
48. Wash'n Ast. & Met. Obs'ns, 1862.
49. " "

CATALOGUE V.—Sporadic Meteors—Continued.

Ref. number	Year	Month	Day	Hour and minute (h. m.)	Place of observation	Apparent size	Color	Duration	Position, altitude and azimuth.
50	1862	Dec.	16	9 30	U.S. Naval Observatory			3(s)	
51	1862	Dec.	15	13 45	U.S. Naval Observatory			5	
52	1862	Dec.	21	11 25	U.S. Naval Observatory			5	
53	1862	Dec.	20	12 35	U.S. Naval Observatory				
54	1863	Jan.	11	11 18	U.S. Naval Observatory			5	
55	1863	April	19	12 0	Philadelphia, Pa.			5	
56	1863	April	8	7 50	U.S. Naval Observatory	"Rather less than the moon"		3 to 6	
57	1864	July	8	9 0	Boston, Mass.				Appeared just below Delphinus.
58	1864	July	13	10 15	U.S. Naval Observatory	Large as Vega.			Appeared 3° S. of Mars.
59	1864	Oct.	29	9 43	U.S. Naval Observatory	More brilliant than Jupiter.			Appeared 7° W. of the zenith.
60	1864	Nov.	19	7 0	U.S. Naval Observatory				In the N.E.
61	1865	Jan.	21	12 0	U.S. Naval Observatory				In the N.
62	1865	Feb.	20	12 59	U.S. Naval Observatory				
63	1865	Nov.	26	8 15	U.S. Naval Observatory		Red		In Cygnus.
64	1866	July	26	8 15	Hector, N.Y.			½ to 1	Seen to explode in N.N.E. at alt. of 10 or 14°
65	1866	Nov.	15	6 15	Glen Gove, N.Y.				In the W.
66	1867	Mar.	29	12 0	U.S. Naval Observatory				
67	1867	April	7	9 5	U.S. Naval Observatory				
68	1867	July	18	7 30	Salem, Pa.	One-quarter diam. of moon.		4 to 5	Appeared in the S.E., disappeared in the N.E. 30° above eastern horizon.
69	1867	Oct.	15	5 40	Haverford College, Pa.	6' in diameter		5	Appeared 3° N. of Pleiades.
70	1868	Feb.	6	11 33	U.S. Naval Observatory	18' in diameter.	Pink		Appeared in W. at alt. 55°, disappeared at alt. 12°.
71	1868	June	6	11 40	Manhattan, Kan.				
72	1869	Jan.	20	10 30	U.S. Naval Observatory	"Of unusual brilliancy"		5	Appeared alt. 25°; az., N. 29° W.
73	1869	May	29	11 20	New York, N.Y.				Appeared in E.; alt. 39°; disappeared ½ 60° E.
74	1869	Aug.	24	7 17	U.S. Naval Observatory				
75	1869	Oct.	26	15 0	Patterson, Ky.	"As large as a basket."			From Mizar to 10° below Polaris.

50. Wash'n Ast. & Met. Obs'ns, 1862.
51. "
52. "
53. "
54. Wash'n Ast. & Met. Obs'ns, 1863.
55. "
56. A. J. S. XXXVI., 154.
57. Wash'n Ast. & Met. Obs'ns, 1864.
58. A. J. S. XXXVIII., 226.
59. Wash'n Ast. & Met. Obs'ns, 1864.
60. "
61. Wash'n Ast. & Met. Obs'ns, 1865.

CATALOGUE V.—Sporadic Meteors—Continued.

Ref. number.	Train, appearance and duration.	Length of path.	Direction of motion.	Remarks.	Observer.
50		°	From zenith to E.		
51			From the N. E.		
52			From zenith to N. W.		
53	Train visible 5s		From N. E. to N.		
54	Visible train		From E. to N. W.		
55	Visible train	45	From the S. E.		
56			From W. to E.	Disappeared with an explosion.	
57	Train 18° long		From the E. N. E.	Seen also at Westtown and at Wilmington and Odessa, Del.	John Gardner.
58	Train 3 or 4 degrees long		From Delphinus towards Pegasus.	Disappeared with a loud report.	
59	Visible train	40	Obliquely towards horizon.		
60			Vertically to horizon.	Very large and brilliant.	
61				Brilliant.	
62				Passed below some cirro-stratus clouds	David Trowbridge.
63			Towards the N.	Report of explosion heard in 3½m	Isaac Coles.
64	A beautiful blue train	39	From E. to W.		
65				Brilliant.	
66				When first seen its altitude was 40°	
67				Disappeared with an explosion.	
68	Train 3 or 4° long		Vertically to the horizon.	Mag. at appearance — η Tauri, at disappearance — Jupiter.	J. K. Larimore.
69		25	Towards horizon at an angle of 75°.	Visible less than 1s. Disappeared with a report heard over	Pliny Earle Chase.
70	Train visible 1m		Vertically to horizon.	an area 120 miles in diameter.	Prof. B. V. Mudge.
71					
72	Brilliant train	39		Burst into fragments with a loud explosion. Seen in many places in New England, New York, and New Jersey.	Described by Prof. E. Loomis.
73					
74	Train visible 5m.			Seen by many observers.	Described by Prof. J. L. Smith.
75	Train 30 feet long, 3 feet wide				

62. Wash'n Ast. & Met. Obs'ns, 1865.
63. "   "
64. A. J. S. XLII₂, 286.
65. Smithsonian Archives (not yet printed).
66. Wash'n Ast. & Met. Obs'ns, 1867.
67. Wash'n Ast. & Met. Obs'ns, 1867.
68. A. J. S. XLIV₂, 288.
69. Proc. Am. Phil. Soc. X, 353.
70. Wash'n Ast. & Met. Obs'ns, 1868.
71. A. J. S. XLVI₂, 429.
72. Wash'n Ast. & Met. Obs'ns, 1869.
73. A. J. S. XLVII₂, 145.
74. Wash'n Ast. & Met. Obs'ns, 1869, 372; Proc. Am. Phil. Soc. XI, 194.
75. A. J. S. XLIX₂, 139.

CATALOGUE V.—Sporadic Meteors—Continued.

Ref. number	Year	Month	Day	Hour and minute	Place of observation	Apparent size	Color	Duration	Position, altitude and azimuth.
76	1869	Nov.	19	6 53	U.S. Naval Observatory	Very bright	------	4.	In the N. W.
77	1869	Dec.	24	8 30	U.S. Naval Observatory	------	------	4	In the E.
78	1870	Mar.	25	13 20	U.S. Naval Observatory	3d magnitude	White	0.4	
79	1870	Mar.	25	8 45	U.S. Naval Observatory	4th magnitude	White	0.3	
80	1870	April	12	8 36	U.S. Naval Observatory	3d magnitude	White	2.4	
81	1870	April	19	8 57	U.S. Naval Observatory	4th magnitude	White	0.5	
82	1870	April	19	9 33	U.S. Naval Observatory	3d magnitude	White	0.5	
83	1870	April	19	10 12	U.S. Naval Observatory	4th magnitude	White	0.5	
84	1870	April	19	15 6	U.S. Naval Observatory	3d magnitude	White	0.6	
85	1870	April	25	11 11	U.S. Naval Observatory	4th magnitude	White	0.6	
86	1870	April	27	11 3	U.S. Naval Observatory	4th magnitude	White	0.6	
87	1870	June	19	14 12	U.S. Naval Observatory	5th magnitude	Reddish	0.4	
88	1870	June	21	10 0	U.S. Naval Observatory	2d magnitude	Reddish	0.3	
89	1870	June	24	10 35	U.S. Naval Observatory	5th magnitude	White	2.0	
90	1870	June	24	10 23	U.S. Naval Observatory	6th magnitude	Orange	0.4	
91	1870	June	25	12 36	U.S. Naval Observatory	1st magnitude	Orange and green	1.5	Midway between Spica and Antares.
92	1870	June	25	12 36	U.S. Naval Observatory	3d magnitude	White	2.5	
93	1870	June	26	9 56	U.S. Naval Observatory	6th magnitude	Orange and green	0.3	
94	1870	July	26	9 50	U.S. Naval Observatory	Size of Jupiter	White	0.2	From 200° +73° to 196° +73°
95	1870	July	30	8 59	U.S. Naval Observatory	Size of Jupiter	White	0.2	From 14h. 50m. + 20°.
96	1870	July	30	9 14	U.S. Naval Observatory	3d magnitude	White	3.0	From 17h. 0m. — 25°.
97	1870	July	30	9 20	U.S. Naval Observatory	Size of Saturn	White	1.8	From 4° south of Saturn.
98	1870	July	30	9 40	U.S. Naval Observatory	3d magnitude	White	0.2	From α Aquilæ.
99	1870	July	30	9 50	U.S. Naval Observatory	4th magnitude	White	0.5	From α Aquarii.
100	1870	Aug.	15	9 54	U.S. Naval Observatory	3d magnitude	Orange	0.5	From α Pegasi.
101	1870	Sept.	4	8 5	U.S. Naval Observatory	4th magnitude	White	0.4	From 10° N. of Arcturus.
102	1870	Sept.	24	8 15	U.S. Naval Observatory	4th magnitude	White	0.4	
103	1870	Sept.	24	8 30	U.S. Naval Observatory	4th magnitude	White	0.3	
104	1870	Oct.	15	8 10	U.S. Naval Observatory	3d magnitude	White	6.3	From 19h. 30m. + 63° directly tow'ds Polaris.

76. Wash'n Ast. & Met. Obs'ns, 1869.
77. "
78. Wash'n Ast. & Met. Obs'ns, 1870.
79. "
80. "

81. Wash'n Ast. & Met. Obs'ns, 1870.
82. "
83. "
84. "
85. "

86. Wash'n Ast. & Met. Obs'ns, 1870.
87. "
88. "
89. "
90. "

CATALOGUE V.—Sporadic Meteors—Continued.

Ref. number.	Train, appearance and duration.	Length of path.	Direction of motion.	Remarks.	Observer.
76	No train	8	N. 20° E	From δ Aquilæ	Prof. J. R. Eastman.
77	No train	5	S. to N	Across the "Sickle"	Prof. J. R. Eastman.
78	Short train	10	S. 5° W	From a point on the meridian 10° S	Prof. J. R. Eastman.
79	No train	4	S. W	South of a Canis Majoris	Prof. J. R. Eastman.
80	No train			Northerly from near the eastern horizon	N. Cahill.
81	No train			Northerly from near the eastern horizon	N. Cahill.
82	No train			Northerly from near the eastern horizon	N. Cahill.
83	No train			Across the "Dipper"	N. Cahill.
84	No train	10	Towards the W	From 16° south of the zenith	D. Horigan.
85	No train	10	Towards the E		D. Horigan.
86	No train	8	N.W. from a Cygni		D. Horigan.
87	Long train, lasting 3s.	12	N. from a Boötis		D. Horigan.
88	No train	6		From ζ Ursæ Minoris to 8° south of a Boötis	Prof. J. R. Eastman.
89	No train	10	South	From alt. 40° through the constellation of the Scorpion	Prof. J. R. Eastman.
90	No train	8	Easterly	From β Cassiopeæ to 16° below Polaris	Prof. J. R. Eastman.
91	Train visible 0.8s	30	West	Exactly across Polaris in a line parallel to the above course	Prof. J. R. Eastman.
92	No train	4			Prof. J. R. Eastman.
93	No train	45	East		Prof. W. Harkness.
94	No train	6	West		Prof. J. R. Eastman.
95	No train	15	West		Prof. J. R. Eastman.
96	No train	15	West		Prof. J. R. Eastman.
97	No train	6	South		Prof. J. R. Eastman.
98	No train			Between 9h. and 10h. saw ten other small meteors, all moving westward.	Prof. J. R. Eastman.
99	No train	8	West		Prof. J. R. Eastman.
100	Train visible 0.6s	10	Southwest	"Across the Dipper"	D. Horigan.
101	No train	8	Vertically to horizon		D. Horigan.
102	No train	8	East	"From Polaris towards the Dipper"	D. Horigan.
103	Train visible 0.3s	10			D. Horigan.
104	No train	4			Prof. J. R. Eastman.

91. Wash'n Ast. & Met. Obs'ns, 1870.   96. Wash'n Ast. & Met. Obs'ns, 1870.   101. Wash'n Ast. & Met. Obs'ns, 1870.
92.   "   97.   "   102.   "
93.   "   98.   "   103.   "
94.   "   99.   "   104.   "
95.   "   100.   "

CATALOGUE V.—Sporadic Meteors—Continued.

Ref. number.	Year	Month	Day	Hour and minute	Place of observation.	Apparent size.	Color.	Dura-tion.	Position, altitude and azimuth.
								s.	
105	1870	Nov.	3	A. m. 11 29	U. S. Naval Observatory	3d magnitude	White	0.3	From ε Andromedæ.
106	1870	Nov.	14	11 59	U. S. Naval Observatory	3d magnitude	White	0.5	From Orion towards Sirius.
107	1870	Nov.	15	7 18	U. S. Naval Observatory	1st magnitude	White	2.0	
108	1870	Nov.	15	7 25	U. S. Naval Observatory	3d magnitude	White	0.5	
109	1870	Dec.	11	9 5	U. S. Naval Observatory	4th magnitude	White	2.0	
110	1870	Dec.	13	9 30	U. S. Naval Observatory	3d magnitude	White	0.2	
111	1870	Dec.	13	12 15	U. S. Naval Observatory	3d magnitude	White.		
112	1870	Dec.	23	13 36	U. S. Naval Observatory	1st magnitude	Orange	3.0	From 3° N. of Jupiter.
113	1871	May	23	9 43	U. S. Naval Observatory	1st magnitude	White	3.5	On the meridian, Z. D. S. 85°.
114	1871	July	6	8 35	U. S. Naval Observatory	2d magnitude	White	1.0	8° N. of Antares.
115	1871	July	6	8 38	U. S. Naval Observatory	2d magnitude	White	1.5	
116	1871	July	6	8 45	U. S. Naval Observatory	4th magnitude	White	0.3	
117	1871	July	11	8 55	U. S. Naval Observatory	1st magnitude	White	3.0	15° N. W. of Antares.
118	1871	July	19	8 or 9	Wilmington, N. C.	"Large ball of fire"	Crimson.		
119	1871	Aug.	8	11 45	U. S. Naval Observatory	2d magnitude	White	1.0	4° N. of Algol.
120	1871	Aug.	9	10 45	U. S. Naval Observatory	2d magnitude	White	0.3	Near β Pegasi.
121	1871	Aug.	12	9 53	U. S. Naval Observatory	1st magnitude	White and orange	1.0	At α Aquilæ.
122	1871	Aug.	16	11 36	U. S. Naval Observatory	As large as Sirius			At α Bootis.
123	1871	Aug.	16	8 43	U. S. Naval Observatory	Larger than Jupiter	White		From Aldebaran passed 5° below the Pleiades.
124	1871	Nov.	16	13 33	U. S. Naval Observatory	1st magnitude	White		
125	1871	Jan.	22	8 15	City of Mexico	1st magnitude	Red, then white.	10m.	From β Cassiopeæ towards Polaris.
126	1872	July	30	9 5	U. S. Naval Observatory	1st magnitude	Orange	4	
127	1872	July	28	8 55	U. S. Naval Observatory	1st magnitude	White	1	
128	1872	July	28	9 18	U. S. Naval Observatory	2d magnitude	White	1	
129	1872	Aug.	3	10 20	U. S. Naval Observatory	2d magnitude	White	1	28° E. of Polaris.
130	1872	Aug.	6	9 53	U. S. Naval Observatory	2d magnitude	White	1	At Polaris.
131	1872	Aug.	8	8 18	U. S. Naval Observatory	2d magnitude	Orange and white	2	
132	1872	Aug.	9	8 37	U. S. Naval Observatory	2d magnitude	Orange and white	1.5	
133	1872	Aug.	9	8 40	U. S. Naval Observatory	3d magnitude	White	1	

105, Wash'n Ast. & Met. Obs'ns, 1870.
106, "
107, "
108, "
109,

110, Wash'n Ast. & Met. Obs'ns, 1870.
111, "
112, "
113, Wash'n Ast. & Met. Obs'ns, 1871.
114,

115, Wash'n Ast. & Met. Obs'ns, 1871.
116, "
117, "
118, A. J. S., II s. 227.
119, Wash'n Ast. & Met. Obs'ns, 1871.

CATALOGUE V.—Sporadic Meteors—Continued.

Ref. number	Train, appearance and duration	Length of path	Direction of motion	Remarks	Observer
105	Train visible 0.4s	5	Southwest		D. Horigan.
106	Train visible 0.3s	8	West		D. Horigan.
107	Train visible 2.0s	25		Passed south of α Pegasi	D. Horigan.
108	No train	8		South from ε Cygni	D. Horigan.
109	No train	4			N. Cahill.
110	Train visible 0.3s	22			N. Cahill.
111					N. Cahill.
112	Short train	8	Westerly		N. Cahill.
113		15	Westerly		Prof. J. R. Eastman.
114		15	Westerly		Prof. J. R. Eastman.
115		18	West from Antares		Prof. J. R. Eastman.
116		10	South		Prof. J. R. Eastman.
117		35	West from ε Cassiopeæ		Prof. J. R. Eastman.
118			From South to North		E. S. Martin.
119	Short train, visible 7s	12	From α to β Aurigæ	Very brilliant	D. Horigan.
120		8	South	Burst into three fragments with a loud explosion	Prof. J. R. Eastman.
121	Train visible 2s	15	South		D. Horigan.
122		20	South	Very brilliant	Prof. J. R. Eastman.
123		35	South		Prof. J. R. Eastman.
124	Train visible 1s	25	South	Very brilliant	Prof. A. Hall.
125			From West to East	Seen at several other cities in Mexico	José de la Vega.
126	Train visible 3s	20		Motion seemed to be by impulses, but slow	D. Horigan.
127		8	Fell vertically from ε Cassiopeæ		D. Horigan.
128	Short train	12	South		D. Horigan.
129	Short train	15	Towards γ Ursæ Majoris		D. Horigan.
130	Short train	8	N. E. from Saturn		D. Horigan.
131	Long train	15	S. from ε Cassiopeæ		D. Horigan.
132	Long train	20	S. from β Ursæ Majoris		D. Horigan.
133	Short train	10			D. Horigan.

120. Wash'n Ast. & Met. Obs'ns, 1871.
121. " " "
122. " " "
123. " " "
124. " " "

125. A. J. S. III, 235.
126. Wash'n Ast. & Met. Obs'ns, 1872.
127. " " "
128. " " "
129. " " "

130. Wash'n Ast. & Met. Obs'ns, 1872.
131. " " "
132. " " "
133. " " "

CATALOGUE V.—Sporadic Meteors—Continued.

Ref. number	Date Year	Month	Day	Hour and minute h. m.	Place of observation	Apparent size	Color	Duration	Position, altitude and azimuth
134	1872	Aug.	9	9 5	U. S. Naval Observatory	3d magnitude	White	s.	
135	1872	Aug.	9	9 16	U. S. Naval Observatory	3d magnitude	Orange	1	
136	1872	Aug.	11	9 58	U. S. Naval Observatory	3d magnitude	Yellow-red	1	Appeared in the cluster in sword handle of Perseus.
137	1872	Nov.	12	9 0	U. S. Naval Observatory	3d magnitude	Orange	1.5	From 3° E. of α Orionis vertically towards the horizon.
138	1872	Dec.	9	10 45	U. S. Naval Observatory	1st magnitude	White	1.9	From ζ Ursæ Minoris towards ζ Draconis.
139	1872	Dec.	12	4 53	Bloomington, Ind.				Altitude 40°, 19° E. of meridian.
140	1872	Dec.	12		Louisville, Ky.	"Large"	Red		Appeared in the zenith.
141	1873	Jan.	3	9 0	U. S. Naval Observatory	1st magnitude	White	1	Appeared at γ Orionis.
142	1873	Feb.	14	6 0	New Haven, Conn.		One green; the following yellow.		
143	1873	April	3	11 5	U. S. Naval Observatory	1st magnitude	White	1.5	Appeared near α Virginis.
144	1873	May	15	8 5	Harbor Grace, Newfoundland.				W. S. W. at an altitude of 40°.
145	1873	June	6	9 36	U. S. Naval Observatory	1st magnitude	Orange	1.5	From Polaris vertically towards horizon.
146	1873	June	6	9 37	U. S. Naval Observatory	1st magnitude	Orange	2.0	From Polaris 6° towards δ Ursæ Majoris.
147	1873	Aug.	5	12 17	U. S. Naval Observatory	1st magnitude	Orange	1.5	From ζ Ursæ Minoris south at an angle of 30° with the horizon.
148	1873	Aug.	10	8 10	U. S. Naval Observatory	1st magnitude	Orange	3.0	Appeared at θ Cygni.
149	1873	Nov.	24	6 0	U. S. Naval Observatory	Bright as new moon	Intense white	2.0	Appeared at θ Aquilæ.
150	1873	Dec.	14	7 30	U. S. Naval Observatory	3d magnitude	White	4.0	From 12° E. of Polaris vertically towards the horizon.
151	1873	Dec.	24	7 39	Washington, D. C.			3	Entered the atmosphere over the State of Delaware at a height of 90 miles, and disappeared over Fairfax Co., Va., at a height of from 8 to 20 miles.
152	1874	May	25	9 13	U. S. Naval Observatory	1st magnitude	Orange and violet	3	Appeared near Polaris.
153	1874	July	18	10 0	Louisville, Ky.	One-third diam. of moon.	Greenish white		
154	1874	Aug.	19	15 20	U. S. Naval Observatory	18' in diameter	White and orange	3	Appeared 5° above Mars and passed 8° W. of Mars.
155	1875	July	9	9 45	U. S. Naval Observatory	Bright as Mars.		3.5	

134. Wash'n Ast. & Met. Obs'ns, 1872.
135. "   "
136. "   "
137.

138. Wash'n Ast. & Met. Obs'ns, 1872.
139. A. J. S. X, 313.
140. A. J. S. X, 253.
141. Wash'n Ast. & Met. Obs'ns, 1873.

142. A. J. S. V, 318.
143. Wash'n Ast. & Met. Obs'ns, 1873.
144. A. J. S. VI, 154.
145. Wash'n Ast. & Met. Obs'ns, 1873.

CATALOGUE V.—Sporadic Meteors—Continued.

Ref. number.	Train, appearance and duration.	Length of path.	Direction of motion.	Remarks.	Observer.
134	Short train	8	S. from ε Ursa Majoris		D. Horigan.
135		15	S. from a Andromeda		D. Horigan.
136	Orange train	15	To 7° E. of Polaris.		Prof. J. R. Eastman.
137		15	E. of S., making an angle of 60° with the horizon.		Prof. J. R. Eastman.
138					Prof. J. R. Eastman.
139		10		Seen also at several places in Kentucky.	Prof. D. Kirkwood.
140	Left a dense stream of blue smoke, which remained visible several minutes.		Towards the horizon in a southerly direction.	Seen from several other points, and at some places its disappearance was followed by loud detonation. (Probably same meteor as preceding.)	Prof. J. L. Smith.
141	Short train	10	Vertically towards the horizon.		D. Horigan.
142			North and down to an alt. of 18° in the direction N. 64° W.	Two meteors close together. Seen also at New Britain, Conn.	Prof. H. A. Newton.
143		10	Vertically towards the horizon.		D. Horigan.
144	Yellow train	8	To W. N. W. at altitude of 10°.	Burst, leaving a coppery-red cloud visible 30m.	Henry H. Cleft.
145		8			Prof. J. R. Eastman.
146		15			Prof. J. R. Eastman.
147	Short train	15			Prof. J. R. Eastman.
148	Short train	45			Prof. J. R. Eastman.
149	Train separated into two distinct parts.		Across α Aquilæ		A. N. Skinner.
150	No train	15		This meteor was seen by many observers from Delaware to Virginia, and all the available data were collected and discussed by Cleveland Abbe in a report to the Washington Philosophical Society.	Prof. J. R. Eastman.
151					
152	Long train	30	Southeasterly	When within 20° of the horizon it burst into three or four fragments, flashing forth red and blue lights.	D. Horigan.
153			From N. N. W. to S. S. E.	Seen through the slit of the great dome from zenith towards S. S. E.	Prof. J. L. Smith.
154	Continuous train	30			Prof. E. S. Holden.
155	Continuous train	15	Fell vertically.	Separated into two parts, each of which moved 5° before extinction.	D. Horigan.

146. Wash'n Ast. & Met. Obs'ns, 1873.
147. " " " "
148. " " " "
149. " " " "
150. Wash'n Ast. & Met. Obs'ns, 1873.
151. Bull. Wash. Phil. Soc. II, 139.
152. Wash'n Ast. & Met. Obs'ns, 1874.
153. A., J., S., Xs, 203.
154. Wash'n Ast. & Met. Obs'ns, 1874.
155. Wash'n Ast. & Met. Obs'ns, 1875.

CATALOGUE V.—Sporadic Meteors—Continued.

Ref. number	Year	Month	Day	Hour and minute h. m.	Place of observation	Apparent size.	Color.	Duration.	Position, altitude and azimuth.
156	1875	July	14	9 10	U. S. Naval Observatory	Large as Mars	Orange and red	4	Appeared 10° above Mars and passed 6° E. of Mars.
157	1875	Aug.	5	11 15	U. S. Naval Observatory	2d magnitude	Orange	2	Moved from λ Sagittarii towards λ Scorpii.
158	1875	Sept.	16	7 36	U. S. Naval Observatory	Three times size of Venus	White	3	Moved from Saturn to λ Aquarii.
159	1875	Oct.	17	7 15	U. S. Naval Observatory	1st magnitude	White and green	2	Appeared near β Aquarii.
160	1875	Nov.	14	13 11	U. S. Naval Observatory	Size of Mars	White	2	Appeared in Cassiopeae.
161	1875	Nov.	17	7 11	U. S. Naval Observatory	3d magnitude	White	1	Appeared near Polaris.
162	1875	Nov.	17	7 20	U. S. Naval Observatory	1st magnitude	White	0.7	Appeared at ζ Ceti.
163	1875	Nov.	17	7 20	U. S. Naval Observatory	1st magnitude	White	0.7	Appeared at χ Ceti.
164	1875	Nov.	17	7 20	U. S. Naval Observatory	1st magnitude	White	0.7	Appeared at χ Ceti.
165	1875	Nov.	22	8 0	U. S. Naval Observatory	8′ to 11′ in diameter	Violet, or pink in front, bright green in center, and pale green following.	8 to 8	Appeared at χ Ceti.
166	1875	Dec.	27	9 0	U. S. Naval Observatory	....	White	Few sec de.	From W. to E. across the northern sky.
167	1876	Jan.	26	9 31	U. S. Naval Observatory	2d magnitude	White	0.7	From γ Canis Majoris to o′ Canis Majoris.
168	1876	Jan.	31	5 30	Louisville, Ky	½ diameter of moon	....	2 or 3	60° above the N. W. horizon.
169	1876	April	19	10 15	U. S. Naval Observatory	3d magnitude	Orange and red	1	
170	1876	July	8	8 45	U. S. Naval Observatory	2d magnitude	White and green	9	Appeared at an altitude of 88 miles.
171	1876	Aug.	9	8 33	U. S. Naval Observatory	2d magnitude	White and green	3	S. W. from ε Urs. Min., towards Arcturus.
172	1876	Aug.	9	11 34	U. S. Naval Observatory	2d magnitude	White and green	3	S. W. from midway between Polaris and Cassiopeae.
173	1876	Aug.	13	11 58	U. S. Naval Observatory	3d magnitude	White	1	Appeared 8° below Polaris.
174	1876	Aug.	14	9 5	U. S. Naval Observatory	3d magnitude	White	1	
175	1876	Aug.	15	9 8	U. S. Naval Observatory	3d magnitude	White	1	
176	1876	Aug.	15	9 9	U. S. Naval Observatory	3d magnitude	White	1	Appeared 5° below ε Cassiopeae.
177	1876	Aug.	15	12 5	U. S. Naval Observatory	3d magnitude	White	1	
178	1876	Aug.	21	8 30	U. S. Naval Observatory	4th magnitude	White	1	
179	1876	Aug.	22	9 40	U. S. Naval Observatory	4th magnitude	White	0.5	
180	1876	Aug.	27	11 25	U. S. Naval Observatory	3d magnitude	White	1	

156. Wash'n Ast. & Met. Obs'ns, 1875.
157. "   "
158. "   "

159. Wash'n Ast. & Met. Obs'ns, 1875.
160. "   "
161. "   "

162. Wash'n Ast. & Met. Obs'ns, 1875.
163. "   "
164. "   "

CATALOGUE V.—Sporadic Meteors—Continued.

Ref. number	Train, appearance and duration.	Length of path.	Direction of motion.	Remarks.	Observer.
156	Continuous train	◇	Fell vertically		D. Horigan.
157	Slight train	12		Exploded and illuminated the small cloud behind which it passed.	Prof. J. K. Eastman.
158	Slight train	18			D. P. Todd.
159	Train visible 20s	25	Fell vertically		D. P. Todd.
160	No train	32	Fell nearly vertically		D. Horigan.
161	Short train	19	Fell vertically		D. P. Todd.
162	No train		Fell vertically		D. P. Todd.
163	No train		Fell vertically	All disappeared at an altitude of 5°.	D. P. Todd.
164	No train		Fell vertically		D. P. Todd.
165	Pale greenish train		Moved vertically through * Cygni and nearly through β Cygni.	Within 6° or 8° of the horizon it passed behind a bank of clouds.	Prof. E. S. Holden.
166	Faint train	8		Seen in Missouri, Kansas, Iowa, Nebraska, and Ohio.	G. C. Brodhead.
167				Seen by several other observers in Kentucky.	D. P. Todd.
168			From N. W. to S. E.		Prof. J. L. Smith.
169	Train short, green.	15	West from Jupiter		D. Horigan.
170	Train visible 40m.		From N. 78° W. over N. E. corner of Indiana.	Exploded over Lake Michigan at an altitude of 34 miles.	Prof. D. Kirkwood.
171	Long train	29			D. Horigan.
172	Train lasted 12s.	30	Westward		D. Horigan.
173	Long train, visible 3s.	15	South from α Andromeda.		D. Horigan.
174	Short train	29	"South from the Dipper".		G. Anderson.
175	Short train	15	Eastward		D. Horigan.
176	Short train	15	West across Polaris		D. Horigan.
177	Long train	29	N. E. from β Pegasi		D. Horigan.
178	Short train	15	N. W. from Polaris.		D. Horigan.
179	Short train	12	E. S. E. from Saturn		D. Horigan.
180	Short train	19			D. Horigan.

165. Wash'n Ast. & Met. Obs'ns, 1875.
166. Trans. St. Louis Acad. Sci. III, 349.
167. Astron'l and Meteor'l Observations, 1876.
168. A. J. S. XI., 488.
169. Wash'n Ast. & Met. Obs'ns, 1876.
170. A. J. S. XIV., 75.

171. Wash'n Ast. & Met. Obs'ns, 1875.
172. "   "   "   "
173. "   "   "   "
174. "   "   "   "
175. "   "   "   "

176. Wash'n Ast. & Met. Obs'ns, 1876.
177. "   "   "   "
178. "   "   "   "
179. "   "   "   "
180. "   "   "   "

CATALOGUE V.—Sporadic Meteors—Continued.

Ref. number	Year	Month	Day	Hour and minute	Place of observation	Apparent size.	Color.	Duration.	Position, altitude and azimuth.
181	1876	Aug.	28	13 31	U. S. Naval Observatory	3d magnitude	White	0.5	Appeared at altitude of 20°; disappeared at altitude of 10°.
182	1876	Oct.	31	10 30	U. S. Naval Observatory	Brighter than Venus	Green	5	Appeared near η Aurigæ.
183	1876	Nov.	5	11 33	U. S. Naval Observatory	3d magnitude	Orange and white	1	Appeared at 10° above and 5°S. of the moon.
184	1876	Nov.	5	11 34	U. S. Naval Observatory	3d magnitude	Orange and white	1	First seen at 12° or 15° N. of W. at an altitude of 10°.
185	1876	Nov.	29	5 13	Bloomington, Ind	Four times size of Venus	White and green	3m.	
186	1876	Dec.	21	8 45	Bloomington, Ind	Larger than the moon			Appeared in the S.E.; disappeared at altitude of 26° or 35°.
187	1877	Feb.	7	14 30	Elletsville, Ind	Half the full moon		0.8	Appeared near π Herculis.
188	1877	Feb.	21	11 29	U. S. Naval Observatory		Yellow	1	Appeared near π Cephio; disappeared midway ψ and γ Cassiopeæ.
189	1877	April	21	9 56	U. S. Naval Observatory	2d magnitude			
190	1877	May	12	8 32	U. S. Naval Observatory	1st magnitude	Orange	3	Appeared in zenith and moved towards x Aurigæ.
191	1877	June	12	8 45	Bloomington, Ind	One-quarter size of moon			Appeared at 18° E. of N., altitude 17° or 18°.
192	1877	July	3	11 10	Bloomington, Ind	Much brighter than Jupiter	White		Appeared at a Ursæ Majoris.
193	1877	Nov.	2	8 49	U. S. Naval Observatory	1st magnitude			From near β Ceti to 5° S. E. of a Piscis Australis.
194	1877	Nov.	11	6 3	Racine, Wis		Yellow, then green	8	Appeared in N. N. E., altitude 30°.
195	1877	Nov.	11	6 36	Osceola, Ark		Golden, then green	19	Appeared in N. E., altitude 30°.
196	1878	June	2	14 59	Chicago, Ill	Equal to moon 4 days old	Red, blue, and violet		Near zenith.
197	1878	June	6	9 25	Cambridge, Mass	1st magnitude			Passed over e Ursæ Majoris.
198	1878	July	22	10 5	Cambridge, Mass	1st magnitude			From 305° +58° to 315° +66°
199	1878	July	23	9 21	Cambridge, Mass	1st magnitude			From 306 +45 to 312 +42
200	1878	July	23	10 4	Cambridge, Mass	1st magnitude			From 315 +9 to 318 —1
201	1878	July	33	10 28	Cambridge, Mass	1st magnitude			From 283 +15 to 286 +11
202	1878	July	27	9 55	Cambridge, Mass	1st magnitude	Blue		From 256 +56 to 252 +38
203	1878	July	28	9 7	Cambridge, Mass	Equal to Mars			From 232 +56 to 6 +51
204	1878	July	28	10 21	Cambridge, Mass	1st magnitude	Deep red	3	From 339 +23 to 347 +39
205	1878	July	28	10 28	Cambridge, Mass	Larger than 1st magnitude			From 316 +13 to 336 +32

181. Wash'n Ast. & Met. Obs'ns, 1876.   183. Wash'n Ast. & Met. Obs'ns, 1876.   185. Wash'n Ast. & Met. Obs'ns, 1876.
182. "   184. "   186. A. J. S. XIII., 168, 243, 277; XIV., 219.

CATALOGUE V.—Sporadic Meteors—Continued.

Ref. number.	Train, appearance and duration.	Length of path.	Direction of motion.	Remarks.	Observer.
181	Short train		West from Polaris		D. Horigan.
182	Train of short duration	10 or 15	From N. towards N. E.		Prof. E. S. Holden.
183		8	Fell vertically to horizon		D. Horigan.
184		8	Fell vertically to horizon		D. Horigan.
185	Train 1.8° long, lasted 1s	15	Down and north	Like the preceding, but 10° further north	D. P. Todd.
186	Followed by a great number of small meteors		From N. W. to N. E.	Passed within 3° of moon's south limb; moon nearly full. Seen in Kansas, Missouri, Illinois, Indiana, and Ohio; disappeared with an explosion.	Prof. D. Kirkwood.
187	Train visible several seconds	4	Westerly	3d magnitude at appearance and three times magnitude of Sirius at disappearance.	J. S. Hunter.
188					D. P. Todd.
189	Light train	15		Disappeared at altitude of 12° or 13°	D. P. Todd.
190		30	Eastward		Prof. J. R. Eastman.
191		20	Vertically downwards		F. M. Parker.
192					Prof. E. S. Holden.
193			Towards the west at an angle of 65° with the vertical.	Observer thought he heard the sound of the explosion.	A. N. Skinner.
194		10 to 12			Robert C. Hindley.
195		12 to 15	Fell nearly vertical	Probably same as preceding	Dr. F. L. James.
196			To 4° above β Cassiopeæ	Near a Cassiopea It broke into seven or eight fragments	E. Colbert.
197	A bright train			Rapid	G. W. E. Trouvelot.
198		11+		Quite slow	E. F. Sawyer.
199		6	Across α Cygni	Rapid	E. F. Sawyer.
200		10		Slow	E. F. Sawyer.
201		5	Across ε and ζ Aquilæ	Rapid, probably a Perseid	E. F. Sawyer.
202		18			E. F. Sawyer.
203	Red streak	21		Rapid	E. F. Sawyer.
204		9		Quite slow	E. F. Sawyer.
205		23+			E. F. Sawyer.

187. Proc. Am. Phil. Soc. XVI, 505.
188. Wash'n Ast. & Met. Obs'ns, 1877.
189. "
190. "
191. A. J. S. XIV, 163.
192. Wash'n Ast. w Met. Obs'ns, 1877.
193. "

194. Proc. Am. Phil. Soc. XVII, 340.'
195. "
196. Proc. Am. Phil. Soc. XVIII, 299.
197. Proc. Am. Phil. Soc. XVIb, 299.
198. A. J. S. XVIb, 348.
199. "

200. A. J. S. XVIb, 348.
201. "
202. "
203. "
204. "
205. "

CATALOGUE V.—Sporadic Meteors—Continued.

Ref. number	Year	Month	Day	Hour and minute	Place of observation	Apparent size	Color	Duration (s.)	Position, altitude and azimuth.
206	1878	Aug.	3	9 49	Cambridge, Mass.	1st magnitude			From 340° +24° to 344° +33°
207	1878	Aug.	3	11 22	Cambridge, Mass.	1st magnitude			From 19 +30 to 22 +47
208	1878	Aug.	3	11 27	Cambridge, Mass.	1st magnitude			From 47 +53 to 59 +59
209	1878	Aug.	11	10 10	Bloomington, Ind.	One-third size of moon		2	Appeared in the E. at an altitude of 10°.
210	1878	Aug.	20	8 32	Cambridge, Mass.	1st magnitude			From 155° +72½° to 137½ +79°
211	1878	Aug.	22	8 59	Cambridge, Mass.	1st magnitude			From 185 +79 to 170 +68
212	1878	Aug.	25	10 2	Cambridge, Mass.	Equal to Mars		3	From 218 +63 to 239 +77
213	1878	Aug.	26	8 59	Cambridge, Mass.	1st magnitude	Deep orange		From 269½ +39 to 256½ +27
214	1878	Aug.	27	10 4	Cambridge, Mass.	1st magnitude			From 263½ +48 to 245 +59½
215	1878	Aug.	27	10 19	Cambridge, Mass.	1st magnitude			From 245 +56 to 229 +58
216	1878	Aug.	29	8 46	Cambridge, Mass.	1st magnitude	Deep orange		From 187½ +43 to 185½ +38½
217	1878	Aug.	30	8 32	Cambridge, Mass.	1st magnitude			From 237 +50 to 235½ +62½
218	1878	Aug.	31	8 35	Cambridge, Mass.	1st magnitude			From 212 +24 to 262½ +26
219	1878	Sept.	16	10 8	Cambridge, Mass.	1st magnitude			From 236 +30 to 232 +23
220	1878	Sept.	16	9 0	Henryville, Clark Co., Ind.	One-fourth or one-fifth diameter of moon.			From γ Ursa Majoris over δ Ursa Majoris.
221	1878	Sept.	18	8 29	Cambridge, Mass.	1st magnitude	Orange	2	From 335° -13° to 325° -29°
222	1878	Sept.	19	9 7	Cambridge, Mass.	1st magnitude			From 339 -2 to 326 +16
223	1878	Sept.	22	9 19	Cambridge, Mass.	1st magnitude			From 299 +15 to 295 +2
224	1878	Sept.	21	9 33	Cambridge, Mass.	Larger than Jupiter	Deep orange	2	From 236½ +17½ to 5 +22
225	1878	Sept.	23	7 18	Cambridge, Mass.	1st magnitude			From 315½ +12½ to 346 +6¾
226	1878	Sept.	23	9 35	Cambridge, Mass.	Larger than 1st mag.	Green	2	From 30 +28 to 10 +26
227	1878	Sept.	27	9 26	Cambridge, Mass.	1st magnitude	Blue	2	From 42 +56 to 105 +63
228	1878	Sept.	27	9 10	Cambridge, Mass.	1st magnitude	Orange	1.5	From 341½ +16 to 347 +22
229	1878	Sept.	28	9 52	Cambridge, Mass.	1st magnitude			From 49 +8 to 47 +2
230	1878	Sept.	29	9 5	Cambridge, Mass.	1st magnitude		1	From 15 +3 to stationary.
231	1878	Sept.	28	9 56	Boston, Mass.	Nearly as bright as Mars.	Orange		From 48 +21 to 35 +17½
232	1878	Oct.	28	9 44	Boston, Mass.	Two-thirds diam. of moon.		10	From 31½ +37 to 20 +37
233	1878	Nov.	12	7 0	Washington, Ind.	As brilliant as Sirius.			From α Lyræ to 20° N. W. of Jupiter.
234	1878	Nov.	12	7 15	Boston, Mass.				Through the trapezium in Ursa Minor.
235	1878	Nov.	13	6 40	New Haven, Conn.	1st or 2d magnitude.		1 to 1.5	From α Lyræ to near χ Ophiuchi.

296. A. J. S. XVI, 348. 299. Proc. Am. Phil. Soc. XVIII, 240. 296. Proc. Am. Phil. Soc. XVIII, 241.
297.   "   219 to 219. A. J. S. XVI, 348.   221 to 231. A. J. S. XVI, 348.
298. A. J. S. XVI, 318.

CATALOGUE V.—Sporadic Meteors—Continued.

Ref. number.	Train, appearance and duration.	Length of path.	Direction of motion.	Remarks.	Observer.
206		10	Between η and β Pegasi	Quite rapid	E. F. Sawyer.
207		17	Between β and γ Andromedae	Slow	E. F. Sawyer.
208		3	From γ Persei	Rapid; a Perseid	E. F. Sawyer.
209		20 or 25	Northward	Supposed to have been seen in Oil City and Titusville, Pa.	J. A. Bower.
210		5		Slow	E. F. Sawyer.
211		6		Rapid	E. F. Sawyer.
212	Orange trail	15	From η Herculis	Seen and mapped also by S. C. Chandler, Jr., at Marlboro', N. H	E. F. Sawyer.
213		12		Rapid	E. F. Sawyer.
214		15		Rather slow; an Aug. Lyriad	E. F. Sawyer.
215		10		Rapid; a Lyriad	E. F. Sawyer.
216		6		Rapid	E. F. Sawyer.
217		13		Slow	E. F. Sawyer.
218		9	Near α Bootis	Slow	E. F. Sawyer.
219		8	Near α Cor. Bor.	Rapid	E. F. Sawyer.
220					Benjamin Vail.
221	Orange streak	5		Slow	E. F. Sawyer.
222		14	From near α Aquarii	Rapid	E. F. Sawyer.
223		17		Rapid	E. F. Sawyer.
224	No trail	5	Vertically near β Ceti	Very slow	E. F. Sawyer.
225		10		Rapid	E. F. Sawyer.
226		33			E. F. Sawyer.
227		7		Slow	E. F. Sawyer.
228		10		Very slow	E. F. Sawyer.
229				Rapid	E. F. Sawyer.
230					E. F. Sawyer.
231				Rapid	E. F. Sawyer.
232	Train visible two or three seconds.	8	Westward		E. F. Sawyer.
233				A small meteor branched off just before it reached Ursa Minor	D. E. Hunter.
234					H. E. Stevens.
235					J. J. Skinner.

232. Science Observer, I, 96.
233. Proc. Am. Phil. Soc. XVIII, 241.
234. Science Observer, II, 30.
235.   "   "   "

CATALOGUE V.—Sporadic Meteors—Continued.

Ref. number.	Date			Hour and minute.	Place of observation.	Apparent size.	Color.	Duration.	Position, altitude and azimuth.
	Year.	Month.	Day.	h. m.				s.	
236	1878	Nov.	14	3 30	Hillside Farm, Mass.				South of observer.
237	1878	Dec.	19	9 30	New Haven, Conn.	Very bright		3	From the Crab Nebula to above and to the left of a Ursa Majoris.
238	1878	Dec.	30	7 14	Grand Traverse City, Mich.	Brilliant			
239	1879	Jan.	27	14 28		Larger than moon.			
240	1879	Mar.	14	15 53	Washington, Ind.		Pale blue		S. 10° W.; alt., 25°.

CATALOGUE V.—Sporadic Meteors—Continued.

Ref. number.	Train, appearance and duration.	Length of path.	Direction of motion.	Remarks.	Observer.
		∘ '			
236				Seen during bright sunshine	Thomas Whitaker.
237				Seen in Indiana, Pennsylvania, and Ohio	J. J. Skinner.
238				Explosion heard 4m. after disappearance	Prof. D. Kirkwood.
239			From S. W. to N. E.	Disappeared with an explosion	Thomas T. Bates.
240	Train visible several minutes.				Prof. D. Kirkwood.

236. Proc. Am. Phil. Soc. XVIII, 241.
237. Science Observer, II, 35.
238. Proc. Am. Phil. Soc. XVIII, 241.
239. Proc. Am. Phil. Soc. XVIII, 243.
240. Proc. Am. Phil. Soc. XVIII, 245.

*9 7 8 3 3 3 7 0 6 8 8 8 2 *